Large Language Models

This book serves as an introduction to the science and applications of Large Language Models (LLMs). You'll discover the common thread that drives some of the most revolutionary recent applications of artificial intelligence (AI): from conversational systems like ChatGPT or BARD, to machine translation, summary generation, question answering, and much more.

At the heart of these innovative applications is a powerful and rapidly evolving discipline, natural language processing (NLP). For more than 60 years, research in this science has been focused on enabling machines to efficiently understand and generate human language. The secrets behind these technological advances lie in LLMs, whose power lies in their ability to capture complex patterns and learn contextual representations of language. How do these LLMs work? What are the available models, and how are they evaluated? This book will help you answer these and many other questions. With a technical but accessible introduction:

- You will explore the fascinating world of LLMs, from their foundations to their most powerful applications

- You will learn how to build your own simple applications with some of the LLMs

Designed to guide you step by step, with six chapters combining theory and practice, along with exercises in Python on the Colab platform, you will master the secrets of LLMs and their application in NLP.

From deep neural networks and attention mechanisms, to the most relevant LLMs, such as BERT, GPT-4, LLaMA, Palm-2, and Falcon, this book guides you through the most important achievements in NLP. Not only will you learn the benchmarks used to evaluate the capabilities of these models, but you will also gain the skills to create your own NLP applications. It will be of great value to professionals, researchers, and students within AI, data science, and beyond.

Large Language Models
Concepts, Techniques and Applications

John Atkinson-Abutridy

CRC Press
Taylor & Francis Group
Boca Raton London New York

CRC Press is an imprint of the
Taylor & Francis Group, an **informa** business

Designed cover image: John Atkinson-Abutridy

First edition published 2025
by CRC Press
2385 NW Executive Center Drive, Suite 320, Boca Raton FL 33431

and by CRC Press
4 Park Square, Milton Park, Abingdon, Oxon, OX14 4RN

CRC Press is an imprint of Taylor & Francis Group, LLC

© 2025 John Atkinson-Abutridy

ISBN: 9781032852362 (hbk)
ISBN: 9781032836089 (pbk)
ISBN: 9781003517245 (ebk)

DOI: 10.1201/9781003517245

Typeset in Minion
by codeMantra

To my wife, Ivana, who serves as my constant source of inspiration, motivation, and unwavering support.

Contents

Preface

LANGUAGE STANDS AS OUR most potent tool for communication, allowing us to express ideas and forge connections with others. Throughout the course of human history, language has continually evolved, adapting to the demands and intricacies of our ever-changing society. It has been the focus of study and research for experts in the field of artificial intelligence (AI) worldwide.

In this context, the realm of AI has made remarkable strides in recent years, and large language models exemplify this progress. These models have emerged as captivating and promising innovations, offering us the means to comprehend and enhance our capacity to both generate and decipher language. Within these models, advanced machine learning techniques and natural language processing methodologies converge to produce coherent and contextually relevant text, even when confronted with incomplete or ambiguous input.

This book on large language models aspires to be a comprehensive and pragmatic guide that delves into the theoretical underpinnings and practical applications of this groundbreaking technology. From grasping the intricacies of machine learning algorithms to implementing language models across diverse contexts, this book serves as an indispensable resource for students, researchers, and practitioners with a vested interest in the field of natural language processing.

By perusing the pages of this book, readers will acquire a profound comprehension of the mechanics behind generative language models, along with insights into the latest developments in this perpetually evolving sphere. Furthermore, this book furnishes numerous real-world examples and use cases that underscore the tangible benefits of large language models in everyday life and within a myriad of industries.

This book, *Large Language Models: Concepts, Techniques, and Applications*, offers a comprehensive compass for those seeking to attain an in-depth grasp of large language models. Within its pages, readers will discover an exhaustive elucidation of the fundamental principles underpinning these models, as well as an exposition of the technologies employed in their construction and their extensive applications.

This book caters to both industry professionals and students and researchers with an interest in the realms of AI and natural language processing. As such, readers will encounter a lucid and approachable elucidation of language models, replete with practical examples and case studies designed to illuminate their potential and demonstrate how they can be effectively harnessed in real-world scenarios.

John Atkinson A.

John Atkinson-Abutridy
Santiago (Chile)

Introduction

THIS BOOK SERVES AS an engaging introduction to one of the most captivating and rapidly evolving domains in the field of artificial intelligence: Natural Language Processing (NLP). Specifically, it delves into the world of Large Language Models (LLMs), which empower computers to undertake a wide array of tasks and applications. These include, but are not limited to, machine translation, summary generation, question answering, conversational systems, and document categorization. Within these pages, you will be acquainted with the foundational concepts, the deep learning methodologies that underpin them, cutting-edge LLMs, practical use cases, and contemplations on the future.

In the pages of this book, you'll embark on a journey to comprehend how NLP is revolutionizing our interactions with machines. Witness how they now possess an unprecedented ability to grasp our intent and respond to our inquiries or directives with unparalleled precision.

TARGET AUDIENCE

This book casts a wide net, appealing to a diverse audience that spans both industry and academia:

- **AI professionals and data scientists**: Those immersed in the realm of AI, specifically NLP and deep learning, will find value in the technical underpinnings, algorithms, and techniques employed in LLMs.

- **Students and academic researchers**: Graduate students and researchers who specialize in AI and NLP will discover this book to be an invaluable resource, offering a robust foundation in the concepts and methodologies integral to LLMs.

- **Professionals in related fields**: Individuals engaged in NLP-associated domains such as machine translation, content generation, *chatbot*-driven customer service, or document categorization will gain insights into how LLMs can augment and optimize their work processes.

While this book is designed as an introductory text, familiarity with the following prerequisites will enhance your comprehension:

- Basic machine learning and/or deep learning techniques.
- Proficiency in the Python programming language.

THE STRUCTURE OF THIS BOOK

This book comprises six chapters, each dedicated to a comprehensive exploration of the fundamental principles, techniques, and methodologies underpinning LLMs. One of these chapters offers practical Python exercises designed to apply the language models and concepts introduced in the preceding sections to various NLP tasks. Furthermore, this book provides an extensive bibliography that readers can refer to for a more in-depth understanding of the concepts and techniques discussed within the book.

To enhance reader familiarity and facilitate the search for supplementary resources, this book consistently employs both the basic terminology and its English equivalent (e.g., *encoders and encoders*) throughout. In some instances, the original English term (e.g., *transformers*) is also used for clarity.

The six chapters are structured around the following topics.

CHAPTER 1: INTRODUCTION

This chapter provides an overview of the current state of artificial intelligence, generative models, and language models. It delves into the necessity for and the rapid expansion of LLMs, explores their diverse applications, and contemplates their future implications.

CHAPTER 2: FUNDAMENTALS

Within this chapter, we lay the foundational understanding for NLP. We delve into topics such as representation learning, *embeddings*, basic and sequential neural networks for NLP tasks, *encoder-decoder* models, generative adversarial networks, attention models, and the revolutionary *Transformers* architecture.

CHAPTER 3: LARGE LANGUAGE MODELS

Chapter 3 delves into several prominent LLMs, including but not limited to BERT, GPT, LAMDA, and PaLM. We explore their distinct approaches, architectural designs, methods for pre-training and *fine-tuning*, and the massive *datasets* that fuel their training. Rich with examples, this chapter paints a comprehensive picture of these models.

CHAPTER 4: MODEL EVALUATION

This chapter serves as a guide to evaluating LLMs. It elucidates the key metrics used to assess their efficiency and effectiveness. We also examine the standard *benchmarking datasets* and introduce recent metrics tailored to evaluate regulatory and security dimensions.

CHAPTER 5: APPLICATIONS

Chapter 5 illustrates the practical utility of the language models we've explored, showcasing various real-world applications. These applications span a wide spectrum of NLP tasks, from question answering and semantic search to document categorization, summary generation, *prompt* design, and beyond.

CHAPTER 6: ISSUES AND PERSPECTIVES

The final chapter synthesizes critical considerations and future perspectives regarding the use of LLMs. We delve into ethical concerns, risks associated with their deployment, emerging capabilities, unpredictability, complexity, human alignment, regulatory challenges, and the nuanced balance between their benefits and limitations.

PRACTICE EXERCISES

This book seamlessly blends theoretical concepts with hands-on practical applications. Chapter 6, in particular, offers a variety of exercises that illuminate straightforward examples and real-world applications for a range of NLP tasks, harnessing insights from multiple language models that have been explored throughout this study.

These exercises are presented in Python and are developed within the Google Colab[1] development environment. All source code and test *datasets* referenced in this book are readily available for download at the publisher's website. Alternatively, you may directly request these resources from the author at atkinsonabutridy@gmail.com.

It's worth noting that for certain exercises, access to *Application Programming Interfaces* (APIs) corresponding to specific models (e.g., GPT-3) is required, typically obtained directly from the respective vendor. For instance, you can obtain an API key to access OpenAI's GPT-3 or GPT-4 models from https://platform.openai.com/account/api-keys. Likewise, API keys for select Google services can be found at https://developers.google. com/webmaster-tools/search-console-api/. Once you've acquired any necessary *API key*, simply insert it into the relevant program sections labeled "Insert API key here."

---

[1] https://colab.research.google.com/.

Author Biography

John Atkinson-Abutridy received a PhD in Artificial Intelligence from the University of Edinburgh in Scotland and is currently a full professor at the Faculty of Engineering and Sciences at Universidad Adolfo Ibáñez in Santiago, Chile. Over the years, he has also held full-time academic positions at various Chilean universities and abroad as a visiting professor and researcher at universities and research centers in Europe (France, UK), the Unites States (MIT, IBM, T.J. Watson), and various Latin American universities. Dr. Atkinson-Abutridy's primary research interests span into NLP, textual analytics, artificial intelligence, and bio-inspired computing. His academic career includes the publication of nearly one hundred scientific papers and the authorship of two books. Recently, he has been at the forefront of numerous scientific and technological projects at the national and international levels, serving as an AI consultant for many companies and founding *AI-Empowered*. In recognition of his substantial contributions to the field of computer science, Dr. Atkinson-Abutridy received the Senior Member Award from the Association for Computing Machinery (ACM) in the United States in 2010. Among his notable accomplishments, he pioneered the first worldwide web-based natural language dialogue model in 2005, a precursor to today's ChatGPT system. In 2023, he released the second edition of his book, *Text Analytics: An Introduction to the Science and Applications of Unstructured Information Analysis* (Taylor & Francis, USA), which was recognized as the top choice in the text mining category by the *Book Authority* organization.

## CHAPTER 1

Introduction

1.1 GENERATIVE ARTIFICIAL INTELLIGENCE

In the early stages of artificial intelligence (AI) research, scientists primarily focused on the development of rule-based systems capable of reasoning and decision-making based on predefined sets of rules (Russell, 2020). However, constructing these systems proved to be a laborious task, demanding experts to painstakingly write out the rules. Moreover, these systems were inherently limited, only able to function within the constraints of explicitly programmed rules.

As AI technologies advanced, novel approaches began to emerge, notably machine learning. This paradigm paved the way for various techniques, including the advent of Artificial Neural Networks. These networks empowered computers to autonomously acquire knowledge from extensive sets of pre-labeled training data (Foster, 2019). However, a notable challenge with this approach is its dependency on manually annotated data. This entailed a considerable human effort to assign labels to various forms of data, such as images, text, and audio, to instruct AI systems on what to recognize and process.

Enter generative AI, a paradigm that eliminates the need for labeled data. Generative AI systems achieve this by independently learning from vast *datasets* (Park et al., 2023) and comprehending the inherent relationships within the data, much akin to the way animals learn from their surroundings (Marcus, 2020).

In the realm of generative AI, deep machine learning models take center stage, enabling the creation of new content based on user input, typically in the form of natural language descriptions. This newfound ability extends to generating various types of content, encompassing written text, images, videos, audio, music, and even computer code (Alto, 2023).

To illustrate, when a human inputs a question or statement into a dialogue system or *chatbot* (Adiwardana et al., 2020), like ChatGPT, the system responds by generating a concise yet reasonably detailed written reply. Moreover, users can engage in ongoing conversations with these *chatbots*, entering follow-up questions, with the system retaining earlier details from the conversation.

Generative AI has recently garnered significant attention due to rapid advancements in the field. Notably, OpenAI's ChatGPT[1] has demonstrated the ability to generate text that

DOI: 10.1201/9781003517245-1

is grammatically sound and convincingly human-like. Furthermore, OpenAI's DALL-E[2] tool has made strides in producing realistic images based on natural language instructions. Other tech giants, including Google and Facebook, have also joined the generative AI arena, developing models capable of generating authentic-looking text, images, and even computer programs.

1.1.1 Understanding the Mechanisms of Generative AI

Generative AI operates by generating novel content utilizing a training *dataset*. Researchers feed substantial amounts of data—comprising text, images, music, or other forms of content—into a neural deep learning system known as Generative Adversarial Networks (GAN) (Bengio, 2014). Within this supervised neural network, data is sifted through a system that rewards successes and penalizes errors, gradually learning to recognize and comprehend the intricate relationships within the training data, under the supervision of humans (Babcock and Bali, 2021).

A GAN consists of two neural networks: a generator, responsible for creating new data, and a discriminator, tasked with evaluating the data's authenticity. These two components collaborate in a dynamic process, with the generator continuously refining its outputs based on feedback from the discriminator. The ultimate goal is for the generator to produce content that is virtually indistinguishable from genuine data, as illustrated in Figure 1.1.

For instance, OpenAI's ChatGPT utilizes the Codex system, enabling the translation of natural language descriptions into computer code. This system draws its power from an extensive *dataset* of over 700 GB, compiled from diverse sources including web content, books, magazine articles, websites, technical manuals, emails, song lyrics, plays, scripts, and other publicly accessible resources.

Informally, generative models excel at creating fresh data instances, while discriminative models specialize in distinguishing between various types of data instances. To illustrate, a generative model can produce lifelike images of animals, resembling real creatures, whereas a discriminative model might accurately differentiate between a dog and a cat.

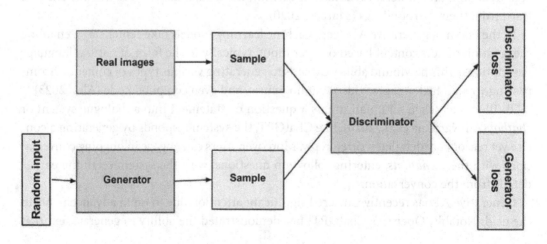

FIGURE 1.1   General architecture of a GAN.

Formally, when presented with a set of X data instances and a set of Y labels:

- A generative model captures the joint probability $p(X, Y)$ or simply $p(X)$ if no labels are involved.

- A discriminative model captures the conditional probability, $p(Y \mid X)$.

Generative models encompass the distribution of the data itself and assess the likelihood of a given example. For instance, models that predict the next word in a sequence, much simpler than GANs, fall into the category of generative models, as they can assign a probability to a sequence of words (Kuhn and Johnson, 2019). Conversely, a discriminative model disregards the likelihood of a given instance or example and focuses solely on estimating the probability of assigning a label to that instance.

Please note that, for instance, in a discriminative classifier like a decision tree, one can assign a label to an instance without necessarily specifying a probability for that label. Nevertheless, such a classifier can still be considered a model, as it effectively models the distribution of all predicted labels to represent the true label distribution within the data. Similarly, a generative model achieves its modeling goal by generating synthetic data that convincingly resembles samples drawn from the target distribution.

Due to this distinction, it's worth highlighting that the modeling task in a generative approach is notably more challenging compared to a discriminative one. For instance, a generative model designed for image data may capture intricate correlations, such as the co-occurrence of boat-like elements near water-like *features* or the absence of eyes on foreheads. Conversely, a discriminative model can discern the difference between sailboats and non-sailboats by recognizing specific discernible patterns, potentially overlooking many of the intricate correlations essential for a generative model.

In essence, discriminative models aim to delineate boundaries within the data space, while generative models strive to understand the arrangement of data points within that space. To illustrate this concept, consider the following general example depicting both discriminative and generative models for a handwritten digit recognition task, as shown in Figure 1.2.

The discriminative model's goal is to differentiate between 0 and 193 handwritten digits based on a linear separation within the data space. When it successfully identifies the correct line, it can distinguish between 0 and 1 without needing an exact model of where the

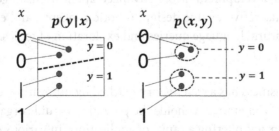

FIGURE 1.2 Discriminative (left) vs. generative models (right).

data instances are positioned on either side of the line. Conversely, the generative model strives to create realistic representations of digits 1 and 39 by generating images that closely resemble their counterparts in the data space. Achieving this requires modeling the distribution across the entire data space. As a result, various image or text generative models have proven effective in training these models to replicate a real distribution. Among these generative models, GANs stand out as a prominent example (Bengio, 2014).

1.1.2 Focus Areas in Generative AI

Generative AI and other AI models are having a profound impact on the advancement of AI technologies, greatly enhancing capabilities for users who may not have a technical background. These innovations encompass the creation of various types of content:

- **Text**: Numerous companies and research laboratories are actively developing natural language interaction capabilities. Examples include Apple's Siri, Google's LaMDA and Bard, Microsoft's Cortana, and Amazon's Alexa. These systems leverage generative AI models to produce written or spoken text.

- **Images**: Generative AI tools like DALL-E and Google's MiP-NeRF have the remarkable ability to generate photorealistic images based on textual input. For instance, a web designer can simply type in "classic Spanish square" into the DALL-E engine and instantly receive a highly realistic image, despite it not representing any real location.

- **Music**: Generative AI extends to the realm of audio and music creation, offering complete compositions and specialized sound effects. Companies like Amper Music, Aiva, Amadeus Code, Google Magenta, and MuseNet are capable of generating original music featuring a variety of lifelike instruments. Users can request specific genres, artists, or styles—be it jazz, Mozart, the Rolling Stones, or an upbeat tempo—and enjoy AI-generated compositions.

- **Software development**: Tools like Amazon's CodeWhisperer and GitHub's CoPilot[3] are already providing natural language-based low-code platforms for developers. Developers can voice or type their queries into these platforms and receive actual lines of software code in various programming languages. This streamlined approach enables developers to work more efficiently and create reusable code modules with ease.

- **Story and game development**: More advanced applications of generative AI involve the creation of narratives, game design, robotic concepts, and even product debugging through natural language queries and exploration of topics.

1.1.3 Applications

The realm of AI in business has a rich history marked by innovation, disruption, and profound transformation. Generative AI holds the potential to guide organizations on a similar transformative journey, offering a range of applications in various domains:

- **Marketing and sales**: Generative AI systems shine in producing diverse forms of content, spanning emails, website text, images, brochures, e-books, product guides, labels, and internal documents. Beyond content creation, these technologies empower organizations to analyze customer feedback, pinpoint both risks and opportunities, and implement highly efficient *chatbots* that enhance customer interactions.

- **Human resources**: HR departments can harness the capabilities of generative AI to craft company handbooks, formulate job descriptions, and devise interview questions. Furthermore, *chatbots* can serve as valuable resources by providing employees with information and self-help options. This might involve streamlining onboarding processes or delivering guidance on choosing the right health insurance or retirement savings plans.

- **Operations**: Generative AI plays a pivotal role in improving customer service through *chatbots* that efficiently handle inquiries, guide individuals to the most pertinent information, and seamlessly transition them to human agents when necessary. Additionally, these systems excel in identifying errors, defects, and other issues through comparative image analysis. For instance, a company can employ generative AI to generate an ideal image of a complex technical component and then employ it to inspect images during the manufacturing process to ensure compliance with quality standards.

- **Software development**: The versatility of generative technology extends to writing code in modern programming languages like Python, Perl, Go, PHP, and JavaScript. Development teams can seamlessly integrate these code snippets and blocks into software projects and store them in libraries for future use. Moreover, generative AI aids in auto-completing data tables and generating synthetic data, thereby enhancing the accuracy of machine learning models. Notably, this technology also proves valuable in simulating cyberattack methods for security assessments.

1.2 GENERATIVE LANGUAGE MODELS

Generative AI has witnessed remarkable advancements in various applications, including image generation systems like Stable Diffusion[4] and DALL-E, alongside natural language dialog systems such as ChatGPT by OpenAI and Bard by Google. These pioneering developments have revolutionized the landscape of AI.

Within the realm of Natural Language Processing (NLP) systems, as extensively covered in works by Eisenstein (2019) and Martin and Jurafsky (2014), the spotlight shines on the use of Large Language Models (LLMs). Notable examples include GPT-3, GPT-4, and LaMDA, as outlined in Zhao et al. (2023). These LLMs empower the generation of fresh text through statistical sampling, leveraging the vast training *datasets* that underpin their creation (see Figure 1.3) (Hu et al., 2023). However, it's worth noting that the roots of generative language models trace back several years, with early explorations aimed at crafting interactive dialog systems based on evolutionary learning from web data (Atkinson, 2005).

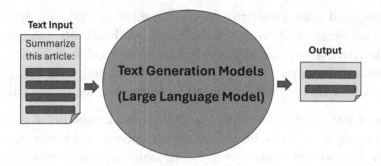

FIGURE 1.3   Role of a large language model.

In essence, an LLM constitutes a neural network boasting an extensive array of parameters, often numbering in the billions, or even more. These models are meticulously trained on copious amounts of unlabeled text through self-supervised learning, as expounded upon in Ge et al. (2023) and Zhao et al. (2023). The overarching goal of LLMs in the realm of generative AI is to assimilate knowledge from textual sources, subsequently generating human-like responses and actively participating in meaningful conversations, as depicted in Figure 1.3 (Bommasani et al., 2021).

The proficiency of LLMs in generating coherent text marks a pivotal advancement in human technology (Gao and Kean, 2023). Notably, these models excel in their capacity to grasp the meaning and context of textual content, encompassing articles, messages, documents, and more, enabling machines to interact with text more intelligently (Wies et al., 2023). However, their true prowess emanates from three distinct facets:

- A single LLM can serve multiple NLP tasks.

- The performance of an LLM scales continually as it accumulates more parameters and undergoes training on larger *datasets*.

- Pre-trained LLMs exhibit the ability to make accurate predictions even when provided with limited labeled examples.

The essence of LLMs (Gao and Kean, 2023) resides in their capacity to acquire optimal *representations* for words and texts, facilitating subsequent manipulation. These mathematical representations, commonly known as *word embeddings* form the cornerstone of various NLP applications, enabling the weighting and classification of different sentences within texts (Martin and Jurafsky, 2014). Conversely, the generative capabilities of LLMs are rooted in their role as predictive neural models, predicting the next word based on prior *embedding* representations. This foundation rests upon a technique called *Recognizing Textual Entailment*, which enhances the comprehension of word relationships. Consequently, as more data is incorporated into LLM training, the system perpetually engages in the analysis of word relationships, seeking *entailment* connections, contradictions, or neutrality.

For instance, let's consider the premise that "a dog has paws." This statement logically *implies* or entails that "paws have feet," but it clearly contradicts the assertion that "dogs swim under the sea." At the same time, it remains neutral regarding an expression like "all dogs are good." As the system processes countless combinations of such statements, it gradually learns to construct accurate and contextually appropriate predictive models. This concept served as the foundation for one of the earliest Language Model with *Transformers* (LLMs), known as BERT, developed by Google.

Typically, LLMs generate text by predicting the most likely words to follow the preceding ones, based on the patterns learned from their training data (Baron, 2019). This training data includes a diverse range of content from the internet, including sources like Wikipedia, as well as materials like fiction, conspiracy theories, propaganda, and more. Consequently, LLMs have the potential to generate content that may be false and/or unverifiable.

In response to these challenges, LLMs began to incorporate human involvement in the training and feedback processes. In applications like ChatGPT, the primary model (such as GPT) was exposed to a *dataset* comprising over three hundred billion words. Initially, human AI trainers played dual roles, simulating both users and AI assistants (either as text generators or evaluators). Subsequently, humans randomly reviewed texts generated by the model, assessed various completions, and provided feedback, thus contributing to the training of a reward model. This iterative process has resulted in the development of a reinforcement learning algorithm known as Reinforcement Learning from Human Feedback. With continued training and user input, the language model continually refines its abilities over time.

The reward predictor evaluates ChatGPT results and predicts a numerical score that represents how well those actions align with the desired system behavior. A human evaluator periodically checks the ChatGPT responses and selects those that best reflect the desired behavior.

As time passes, the *reward* model is updated and refined, producing more realistic results. However, the content generated by LLMs may be biased, unverifiable, constitute original research, and violate copyright; hence, LLMs should not be used for assignments or in subject areas with which the publisher is unfamiliar; hence, their results should be rigorously reviewed for compliance with all applicable policies.

As a consequence, the true autonomy of LLMs depends on the trust and reliability of AI applications, which may emerge as these models improve. For now, humans are the supreme masters, and reliable results depend on collaboration between humans and these AI models.

As a whole, LLMs are game changers for productivity, as they can access and process information in real time, tackle complex problems, and do more in less time; for example, a UK theater group used the GPT-3 model to write a play. The system generated a story based on descriptions from the writers, and the story was further edited before the final version of the narrative was ready for the play. In another case, the Guardian agency used the GPT-3 model to write eight different articles, which were then compiled into one.

1.2.1 Popular Types of LLMs

In the realm of NLP, there exist a diverse array of LLMs known for their exceptional capabilities and performance across different NLP tasks. However, it's crucial to recognize that not all LLMs are created equal. They vary in terms of their design and purpose, with some being general-purpose models, others fine-tuned for specific tasks, and some engineered to operate efficiently on low-capacity devices. These approaches come with distinct strengths, and weaknesses:

- **BERT** (Bidirectional *Encoder* Representations from *Transformers*): It is a pre-trained LLM employing deep learning techniques to generate natural language text. It stands out for its bidirectional approach, considering both left and right context when predicting the next word in a sequence. During training, BERT is exposed to sentences or text sequences with certain words masked, challenging the model to predict the missing words. This process equips BERT to capture the contextual relationships between words within a sentence. It can also be fine-tuned for specific NLP tasks, such as text classification, question answering, and machine translation.

- **GPT-3** (Generative Pre-trained *Transformer*): GPT-3 is an *autoregressive* model, pre-trained on an extensive corpus of text to generate high-quality natural language text. Noteworthy for its adaptability, GPT-3 can be tailored to a wide range of language tasks, including summarization and question answering.

- **GPT-4**: Built on the foundation of GPT-3, GPT-4 is a multimodal LLM capable of processing both image and text inputs and producing text-based outputs. Employing pre-trained *Transformers*, it excels in predicting the next *token*, achieving enhanced performance in factuality measures, and adhering to desired behavior through post-training alignment.

- **LaMDA**: It is a language model optimized for dialogue applications, akin to GPT-4. However, LaMDA's distinct training focus is on capturing the intricacies of open-ended conversations, ensuring high-quality natural language text generation in dialogues.

- **LLaMA**: While smaller in scale compared to GPT-3 and LaMDA, it aims for equivalent performance. As an *autoregressive* language model based on *Transformers*, it undergoes training with a larger number of *token*s, enhancing its performance while maintaining a smaller parameter count.

- **BLOOM**: Developed by Facebook AI, BLOOM is an LLM trained through unsupervised learning. BLOOM excels in generating natural language text with remarkable consistency and fluency. It boasts high performance across a broad spectrum of NLP tasks, including text classification and question answering. A distinguishing *feature* of BLOOM is its ability to grasp intricate linguistic structures and semantic relationships among words, enabling it to produce text that closely mirrors human language.

Recently, unpredictable skills of LLMs have been observed that were not present in simpler models (aka emergent skills). Such skills cannot be predicted by simply extrapolating performance from smaller models. Examples include multi-step arithmetic, college-level test-taking, and decoding the International Phonetic Alphabet, among others.

1.3 CONCLUSIONS

Generative AI has revolutionized the field of AI by enabling computational systems to learn on their own and generate new content without relying on predefined rules or labeled data. Through deep machine learning models, such as GANs, generative AI systems can create authentic-looking written text, images, music, and computer code. These advances have opened up numerous opportunities in a variety of areas, such as *marketing*, business operations, *software* development, and story and game creation. LLMs, such as GPT-4 and BERT, have proven to be especially outstanding at generating coherent text and understanding context. However, it is important to note that the results generated by LLMs must be carefully verified and reviewed, as they may be biased, unverifiable, or violate copyright. As we move forward, collaboration between humans and AI systems becomes critical to ensuring the trust and reliability of generative applications. Generative AI has the potential to change the way we work and interact with technology, and it is expected to continue to evolve and improve in the future.

NOTES

1 https://chat.openai.com/
2 https://openai.com/dall-e-2
3 https://github.com/features/copilot
4 https://stablediffusionweb.com/

## CHAPTER 2

Fundamentals

2.1 INTRODUCTION

Natural Language Processing (NLP) encompasses a spectrum of computationally driven techniques designed to analyze and represent naturally occurring text across various levels of linguistic analysis (Burns, 2019; Atkinson, 2022; Martin and Jurafsky, 2014). These techniques aim to emulate human-like language comprehension, catering to a wide array of tasks and applications (Ghosh and Gunning, 2019). While the surge of interest in NLP has been a recent phenomenon, its roots trace back to the late 1940s.

Within the realm of artificial intelligence (AI), *Language Models* (LMs) have assumed a pivotal role (Li, 2022). At its core, an LM is a model that assigns probabilities to sequences of words, ranging from basic *n-grams* to sophisticated neural LMs. Presently, pre-trained model representations, also known as *embeddings*, are derived from classical LMs established through statistical methods or elementary neural networks.

The fundamental objective of an LM is to generate a *logical continuation* of the text it has encountered, where "logical" refers to what one would expect based on billions of documents composed by humans.

To illustrate, consider the text "John bought a." Imagine sifting through this snippet within the vast corpus of human-generated text and determining the subsequent word(s). This process yields a ranked list of words, accompanied by their associated *probabilities*, as depicted in Figure 2.1.

When the goal is to produce more extensive output, such as an essay, the approach is essentially a repetitive inquiry: "Given the text so far, what should the next word be?" This question is posed iteratively, with each addition of a word. Consequently, at each step, a list

John bought a	car	4.5%
	land	3.2%
	banana	3.5%
	book	3.1%
	house	2.9%

FIGURE 2.1   A ranked list of likely words following a sentence.

    DOI: 10.1201/9781003517245-2

of words with corresponding probabilities is generated. However, the choice of the word to incorporate into the evolving text is not straightforward. While one might assume selecting the word with the highest probability is the logical choice, this can result in monotonous essays devoid of creativity. Conversely, occasionally opting for lower-ranked words injects novelty and intrigue into the composition. The introduction of randomness in this decision-making process implies that using the same input message multiple times will likely yield different essays each time.

To illustrate, consider the outcomes when the model repeatedly selects the word with the highest probability at each step:

```
:
John bought a
John bought a car
John bought a car at
John bought a car at a
John bought a car at a sale
```

However, what occurs when dealing with longer texts? In such instances, the result often manifests as a perplexing and redundant composition. So, what if, rather than consistently opting for the word with the highest probability, one occasionally selects words randomly, regardless of their superiority? Once more, one can compose a text resembling the following:

```
John bought a
John bought a book
John bought a book for
John bought a book for coloring
```

The most logical starting point for pondering the capabilities of an LM is to inquire whether it can fulfill its intended purpose: the modeling of language.

In a formal sense, an LM represents a probability distribution over sequences of *tokens*, such as words. Given a vocabulary set V containing various *tokens*, the LM denoted as "p" assigns a probability, ranging from 0 to 1, to each sequence of *tokens*, represented as $x_1,\ldots,x_L \in \mathcal{V}$:

$$p(x_1,\ldots,x_L)$$

This probability serves as an intuitive measure of the quality of a sequence of *tokens*. For instance, if we consider a vocabulary V with *tokens* {bought, a, John, book}, the LM might assign probabilities like this:

p(John, bought, a, book) = 0.02
p(book, bought, a, John) = 0.01
p(book, a, John, bought) = 0.0001

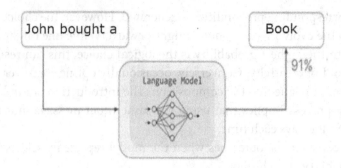

FIGURE 2.2   A simple language model for predicting the following word.

From a mathematical standpoint, an LM appears quite straightforward. However, the capability to assign probabilities to all possible sequences demands exceptional linguistic competence and a profound understanding of the world.

For instance, the LM should implicitly assign a very low probability to the sequence "book a John bought" due to its lack of grammaticality (i.e., syntactic knowledge). Conversely, it should assign a higher probability to "John bought a book" compared to "book bought a John," as illustrated in Figure 2.2. This difference arises from world knowledge; both sentences are syntactically equivalent but vary in their semantic plausibility.

As previously defined, an LM assesses the quality of a given sequence $X_{1:L}$ by returning a corresponding probability. Once we have an LM in place, we can proceed to generate sequences. The most straightforward method to achieve this involves sampling a sequence $X_{1:L}$ from the LM p, with the probability represented as $p(X_{1:})$ This can be denoted as

$$x_{1:L} \sim p$$

The computational efficiency of this process relies on the specific form of the LM p. In practical applications, it's common to directly sample from an LM. This is due to the constraints of real-world LMs and the fact that, at times, our goal is not to obtain an "average" sequence, but rather something closer to the "optimal" sequence.

2.2 *AUTOREGRESSIVE* LANGUAGE MODELS

Traditionally, the representation of the joint distribution $p(X_{1:})$ for a sequence $X_{1:L}$ is established through the chain rule of probability:

$$p(x_{1:L}) = p(x_1)p(x_2 x_1)p(x_3 x_1, x_2) \cdot p(x_L | x_{1:L-1}) = \prod_{i=1}^{L} p(x_i | x_{1:L-1})$$

To illustrate this concept further, let's consider an example:

p(John, bought, a, book) = p(John)
p(bought | John)

$$p(a \mid John, bought)$$
$$p(book \mid John, bought, a)$$

Here, $p(x_i x_{1:i-1})$ represents the conditional probability distribution of the next *token* x_i, given the preceding *tokens* $x_{1:i-1}$. Consequently, an *autoregressive* LM is one where each conditional distribution $p(x_i \mid x_{1:i-1})$ can be efficiently computed, typically utilizing a *feed-forward* neural network.

Now, when generating the complete sequence $X_{1:L}$ from an *autoregressive* LM p, we proceed by sampling one *token* at a time based on the *tokens* generated so far:

$$i = 1, \ldots, L: \qquad x_i \sim p(x_i \mid x_{1:i-1})^{1/T}$$

In this equation, T is a temperature parameter that regulates the level of randomness within the LM:

- $T = 0$ selects the most probable *token*, x_i, at each position i deterministically.

- $T = 1$ performs standard sampling from the pure LM.

- $T = \infty$ uniform samples from the entire vocabulary V.

However, raising probabilities to the power of $1/T$ may result in a distribution that does not sum up to 1. To address this, we normalize the distribution, yielding the "cooled conditional probability distribution." This normalization is defined as $p(x_i \mid x_{1:i-1}) \propto p(x_i \mid x_{1:i-1})^{1/T}$. This concept is reminiscent of the simulated *annealing optimization method*.

For instance,

$$p(book) = 0.4 \qquad p(John) = 0.6$$

$$p_{T=0.5}(book) = 0.31 \qquad p_{T=0.5}(John) = 0.69$$

$$p_{T=0.2}(book) = 0.12 \qquad p_{T=0.2}(John) = 0.88$$

$$p_{T=0}(book) = 0 \qquad p_{T=0}(John) = 1$$

In essence, these models are engineered to enable computers to comprehend, interpret, and generate human language. As a result, LMs have a myriad of practical applications, including *chatbots*, voice assistants, machine translation, and sentiment analysis, among others (Adiwardana et al., 2020).

The evolution of LMs spans several decades, marked by significant transformations. Initially, NLP ventured into rule-based approaches, but these methods quickly revealed their limitations, lacking the necessary flexibility for tackling complex NLP tasks.

With the advent of statistical models, and more recently, deep learning techniques, LMs have undergone a remarkable evolution, becoming substantially more sophisticated and versatile.

2.3 STATISTICAL LANGUAGE MODELS

In the 1990s, a pivotal moment arrived with the emergence of statistical models in NLP. These models heavily relied on vast corpora of textual data to discern intricate language patterns and make predictions concerning novel data (Choi, 2019). This marked a substantial departure from the rule-based systems of the past, sparking a revolution in the NLP field and laying the foundational groundwork for the development of contemporary LMs.

One of the initial statistical approaches in the field of NLP introduced the concept of the hidden Markov model, often abbreviated as HMM (Baron, 2019). HMMs are designed to capture sequences of events that are concealed or unknown within a sentence. Initially, they found their primary application in language modeling, where they are employed to predict the likelihood of a word or label based on the preceding words in a sentence.

Another notable milestone in statistical language modeling was the development of the *n-gram* LM. An *n-gram* represents a sequence of *n* words, and the *n-gram* LM predicts the probability of a word, taking into account the prior *n-1* words. While this model may seem straightforward, it has proven to be remarkably effective and has been widely adopted in numerous NLP tasks, ranging from speech recognition to machine translation.

As the late 1990s approached, NLP began delving into the realm of machine learning techniques. Among these, support vector machines and neural networks emerged as prominent players. These techniques found considerable success in various classification tasks within NLP, such as lexical tagging, also known as *part-of-speech tagging*, and Named-Entity Recognition.

2.4 NEURAL LANGUAGE MODELS

A *Neural Language Model*, often abbreviated as NLM, encompasses the essence of gauging the probability of a word sequence through the utilization of neural networks. This groundbreaking approach extends the concept of acquiring effective *features* for words or sentences, thus birthing a comprehensive neural network methodology to address a wide spectrum of NLP tasks. These pivotal studies ushered in the era of LMs for learning representations, ushering in a profound revolution in the NLP domain (Ekman, 2022).

The inception of the 2000s witnessed the commencement of neural networks being employed for the creation of LMs. This undertaking involves the prediction of the subsequent word in a text, given the context of preceding words. The initial Neural Language Model comprised a *straightforward feed-forward* network housing a hidden layer. It was within this realm that the notion of *distributed word representations*, commonly referred to as word *embeddings*, was first introduced. These word *embeddings* are vectorized, real-valued descriptors that elegantly encapsulate the semantic essence of a word or concept in terms of its distinctive *features* (Peters et al., 2018). The model takes as input vectorized representations of the preceding "n" words relative to the current word. These representations are retrieved from a table that undergoes simultaneous learning with the model. Subsequently, these vectors are fed into a hidden layer, the output of which is subsequently channeled through a SoftMax layer. This architectural design empowers the model to predict the subsequent word within the sequence (Aggarwal, 2018).

In 2013, a pivotal moment in the realm of NLP occurred with the introduction of *Word2Vec*, which quickly rose to prominence as one of the most widely adopted *embedding* models, representing words as vectors. What set Word2Vec apart was its revolutionary approach to training. It achieved this by eliminating the hidden layer and employing an approximation method for the target, effectively streamlining the training process. These seemingly modest alterations paved the way for the scalable training of *embeddings* on extensive text corpora. The training regimen employed by Word2Vec endowed the model with the ability to discern intricate relationships between words. This newfound capacity generated significant intrigue and enthusiasm within the NLP community. These *embeddings* swiftly became a cornerstone of contemporary NLP practices, as their utilization demonstrated marked enhancements in performance across a diverse spectrum of language-related tasks.

In simple terms, Artificial Neural Networks (ANNs) belong to a class of machine learning algorithms inspired by the intricate structure and functionality of the human brain. They comprise layers of interconnected neurons (nodes) with the remarkable ability to learn intricate patterns and intricate data relationships, such as nonlinear connections.

The integration of these techniques has ushered in the development and enhancement of three well-defined categories of neural networks: recurrent neural networks (RNNs), convolutional neural networks (CNNs), and recursive neural networks. RNNs gained initial popularity for handling the dynamic input sequences frequently encountered in NLP. Nevertheless, they were swiftly supplanted by conventional *long short-term memory networks*, also known as LSTMs, as these networks demonstrated greater resilience in mitigating the vanishing gradient problem. Simultaneously, CNNs, which had initially found widespread adoption in computer vision tasks, began making their way into the realm of NLP. The inherent advantage of employing CNNs for processing text sequences is their superior parallelizability compared to RNNs. This stems from the fact that the state at each time step depends solely on the stored context, rather than relying on all preceding states, as is the case with RNNs

Subsequently, the development of sequence-to-sequence, or *Seq2Seq* learning, marked a significant stride in the field of AI. This approach, characterized by its end-to-end nature, employs neural networks to map one sequence onto another. Initially, an encoding neural network, often referred to as the *encoder*, meticulously processes each *token* within a sentence and condenses it into a compact vector representation. Subsequently, a decoding neural network, known as the *decoder*, extrapolates the output sequence *token*-by-*token*. This prediction relies on both the state of the *encoder* and the input of previously forecasted *tokens* at each sequential step. Although the foundational architecture of sequence *encoder-decoder* models predominantly relies on RNNs, more recent innovations have introduced alternatives like Long Short-Term Memory (LSTM) networks and *Transformers* (Raffel et al., 2019; Yue, 2023).

A noteworthy advancement in the realm of NLP is the emergence of *Large Language Models* (LLMs), as exemplified by BERT and GPT. These LLMs, rooted in the *Transformer* architecture introduced by Dai et al. (2019), represent a breakthrough in deep neural networks. They stand as part of a broader class of models known as *foundational models*.

These models have the unique ability to be adapted to a wide array of tasks, thanks to their extensive training on vast amounts of unstructured and unsupervised data (Mialon et al., 2023).

Pre-trained language model (PLM) *embeddings* can serve as *features* within a target model, or alternatively, a pre-trained LM can undergo *fine-tuning* with the target task data. This approach has proven to be highly effective, demonstrating efficient learning even with significantly reduced data volumes. The primary advantage of these pre-trained LMs lies in their capacity to acquire word representations from vast unannotated text corpora. This attribute proves especially advantageous for low-resource languages where the availability of labeled data is scarce.

Furthermore, a significant advancement in the realm of deep learning for NLP has been the integration of *attention mechanisms*. These mechanisms empower models to selectively focus on specific portions of input data during the prediction process. This development has been particularly transformative in the context of sequence-to-sequence models. Among the most influential attention-based models stands the *Transformer* architecture, which has consistently delivered remarkable results. It has played a pivotal role in the creation of some of the most cutting-edge LMs (Kalyan et al., 2021).

Consequently, the emergence of *generative models* (Bengio, 2014; Foster, 2019) has enabled the generation of new data that closely resembles the training data. This innovation has significantly contributed to the development of language models capable of generating human-like text. Notable among these is the GPT-3 model, renowned for its ability to produce text resembling human language. It has found extensive utility in the creation of *chatbots* and other conversational interfaces.

In recent times, we've witnessed remarkable advancements in NLP, particularly with the advent of PLMs like BERT and GPT-3. These models undergo extensive training on massive text *datasets*, enabling them to generate text that mimics human language and comprehend natural language in ways previously considered unattainable, as depicted in Figure 2.3.

2.4.1 Pre-trained Language Models

Among the initial endeavors to generate contextual *embeddings* for words through deep neural networks, the ELMo model stood out. ELMo represents a PLM that employs bidirectional LSTM-type networks rather than relying on fixed word representations. Subsequently, the introduction of BERT marked a significant milestone. BERT is a bidirectional, pre-trained model, trained on vast corpora, utilizing the transformative *Transformers* architecture and attention mechanisms (Vaswani et al., 2017). These pre-trained contextual word representations have proven highly effective as versatile semantic *features*, substantially enhancing the performance of NLP tasks. This breakthrough has spurred a wealth of subsequent research, establishing the paradigm of pre-training and *fine-tuning* in learning. In line with this paradigm, a plethora of studies on PLMs have emerged, introducing various architectures (e.g., GPT-4) and refined pre-training strategies (Lialin et al., 2023).

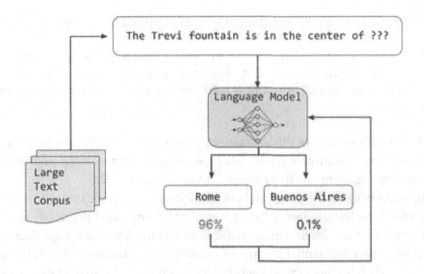

FIGURE 2.3 LM training from large corpus or datasets.

2.5 LARGE LANGUAGE MODELS

Research has consistently demonstrated that scaling up PLMs tends to enhance their performance across various tasks. To explore the limits of model capabilities, recent studies have delved into the realm of larger PLMs, such as the 175-billion-parameter GPT-3 and the massive 540-billion-parameter PaLM. While the primary focus of scaling pertains to model size, retaining similar architectures and task pre-training, these expansive PLMs exhibit distinct behaviors compared to their smaller counterparts, such as the 330-million-parameter BERT and the 1.5-billion-parameter GPT-2. These differences manifest as "emergent abilities," enabling large PLMs to tackle complex tasks. For instance, GPT-3 can adeptly handle *few-shot* tasks through context learning (Wies et al., 2023), a feat that GPT-2 struggles with. One prominent application of LLMs is exemplified in ChatGPT, which leverages GPT-based LLMs to facilitate human-like dialogues and showcases remarkable conversational proficiency.

The success of LLMs largely stems from their capacity to capture intricate word dependencies within text (Mialon et al., 2023). For instance, in the sentence "The cat sat on the mat," the word "cat" depends on the word "the," and the word "mat" depends on the word "on." In an LLM, these dependencies are meticulously encoded within the model parameters. However, it is worth noting that, despite their remarkable advancements, LLMs have witnessed an exponential increase in the number of parameters they employ (Zhao et al., 2023).

2.6 WORD EMBEDDING MODELS

One crucial element in the previously discussed NLP learning approaches lies in the necessity to establish a measure of proximity or similarity between texts or words to effectively address a given task. These tasks can encompass categorizing a document based on its meaning, discerning the emotional tone within an opinion, identifying the resemblance between a user's *query* and potential documents retrievable by a search engine, and more.

To illustrate this concept, let's examine a snippet extracted from a pet-related website:

```
Going for a walk with our furry friend should be a pleasant
experience for both of us, but sometimes it is only for him
because, when he barks at the sight of other animals, it becomes
very annoying.
```

Comprehending the essence of this passage serves as a fundamental input for various NLP tasks. For instance, in sentiment classification, one might inquire, "What emotion does the fragment convey regarding the furry friend?" (Agarwal et al., 2020).

Initially, addressing such a *query* might involve extracting specific syntactic relationships via grammatical analysis, followed by automated reasoning using previously constructed meaning representations for sentences, possibly employing first-order logic. Nevertheless, when dealing with a substantial volume of documents, as demonstrated in the example above, this approach becomes exceedingly intricate, primarily due to the following reasons:

- Manual specification of language rules is necessary, which imposes limitations on both robustness and computational efficiency.

- The task also demands human experts to define and specify basic grammatical rules as well as semantic representation and reasoning mechanisms.

- Due to the inherent high dimensionality of natural language, any method relying solely on these specifications would prove highly inefficient. Constant modifications to linguistic rules would be required, rendering the approach unfeasible.

The example question above shares a common thread with other NLP tasks: the need for meaningful text representation to facilitate subsequent inference. We must address two key questions: how to efficiently represent the meaning of words or sentences within a document, and how to perform operations or inferences on these representations.

One approach to tackling these questions involves the automatic learning of *representations* based on the context in which words appear in texts. Context, in this context, refers to the surrounding environment or window of words that shapes their meaning. A common method for encoding such contexts is to employ straightforward vector representations, such as *word frequency models*, for words and/or documents. Determining proximity becomes relatively straightforward, involving the calculation of distances between these representations. However, the potential number of contexts or *features* can grow exponentially, and not all of them contribute significantly to the representation. Therefore, some form of dimensional reduction typically becomes necessary to efficiently encode representations in a reduced number of dimensions.

To accomplish this goal, computational techniques are essential for converting high-dimensional vector representations into a lower-dimensional model. This transformation allows us to capture concealed or latent relationships between words and documents. The outcome of this process, which generates low-dimensional word vectors,

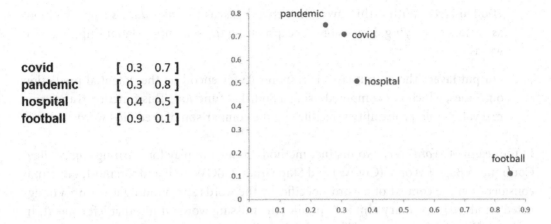

FIGURE 2.4 Vectorial representation in low dimensions (word embeddings).

is commonly referred to as distributed word representations or word *embeddings*. In Figure 2.4, we illustrate the *embeddings* of certain words projected into lower-dimensional spaces. It becomes evident that this transformation unveils previously hidden proximity relationships among words, which, when grouped, seem to reflect genuine linguistic associations more accurately.

Ideally, the learning of these vectors should closely mirror the proximity between words used in similar contexts. The initial method capable of generating these word *embeddings* was known as *Latent Semantic Analysis* (LSA) (Beck, 1988). LSA is an unsupervised learning technique that leverages singular value decomposition to reconstruct an initial vector space of words (or documents) into fewer dimensions, as discussed by Atkinson and Palma (2018). This dimensionality reduction has a significant impact, as it diminishes the influence of dimensions on the importance or weighting of each word within documents. However, one limitation of LSA is its disregard for word context in generating these representations. Additionally, it lacks robustness, often yielding inconsistent results.

Subsequently, more efficient techniques for generating *embeddings* have emerged in the field of AI. These techniques prominently *feature* the use of ANNs, which are trained to predict and reconstruct the contextual relationships of words. This process involves taking a corpus of texts as input and segmenting it into sentences, where words are analyzed within their contextual framework. During training, the neural network learns to predict the appropriate context for words, concurrently *fine-tuning* the optimal *feature* vectors necessary for accurate predictions. As a result, the generated model encompasses low-dimensional vectors derived from the *hidden layers* of the neural network.

One of the most widely adopted models for *embedding* learning belongs to the *Word2Vec* family of methods. To enhance computational efficiency, this approach employs a straightforward neural network model referred to as a three-layered feed-forward neural network (FNN) (Goldberg, 2017). This FNN model comprises the following layers:

1. **Input layer**: This layer contains the encoded word windows extracted from a corpus. The window's size determines the number of words considered within the context.

2. **Hidden layer**: Within this layer, the network learns the *embeddings* from the input as vectors of varying dimensions, encapsulating the semantic relationships between words.

3. **Output layer**: The output layer is responsible for encoding the potential prediction outcomes, which are computed using a SoftMax function. This function transforms real values into probabilities, facilitating the comparison of predicted words.

In the realm of *Word2Vec*, two distinct methods come into play for learning *embeddings*: Continuous Bag of Words (CBOW) and Skip Gram. CBOW, first and foremost, takes into consideration the context of a word, specifically the words surrounding it within a designated window. Its primary aim is to predict the missing word within that left context. In contrast, Skip Gram focuses on the word at the center of attention and endeavors to predict its right-context.

When presented with the context of a word, the CBOW model must adeptly grasp the likelihood that a word (*token*) within the vocabulary serves as a neighboring term to the one initially provided. After successful training of the neural network, it becomes possible to extract hidden layer weight vectors, each of a defined dimension or neurons. These vectors are the very embodiment of the learned *embeddings* for each word within the vocabulary.

For a visual representation of the model's architecture, depicting the transformation from input to output using *Word2Vec*, refer to Figure 2.5:

The training process for this architecture unfolds as follows:

1. The architecture begins with the *creation of two random weight matrices*. The first matrix links the input layer (comprising words from a vocabulary V) to the hidden layer, consisting of N neurons W_{vn} (matrix $V * N$). The second matrix, the context matrix, establishes connections between the N neurons in the hidden layer and the

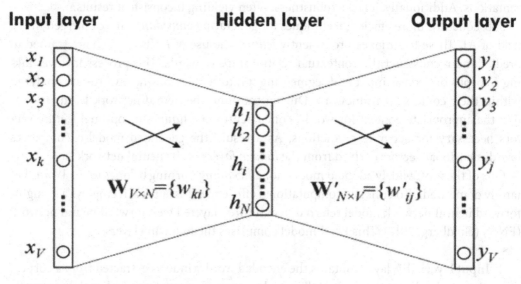

FIGURE 2.5   Word2Vec CBOW model architecture.

output layer W_{nv} (matrix $N * V$). The output is again a vector of length V with the elements (y_i) to predict in the output layer or SoftMax. Both weight matrices have an *embedding* for each vocabulary word when the training is finished.

2. *For each training sample, the following steps are executed:*

a. *Hidden Layer Output Generation*: The hidden layer output is computed by taking a weighted sum of the product of the input vectors and the weight matrix W_{vn}. This product essentially selects the row in the matrix that corresponds to the non-zero elements of a vocabulary word. Consequently, the hidden layer operates as a *lookup table*, producing the word vector associated with the input word. The neurons in the hidden layer simply copy a weighted sum of the inputs to the next layer, bypassing any intermediate activation function.

b. *Output Layer Value Generation*: The output layer of the model comprises another vector of the same length as V and represents the likelihood of each target word being a neighbor of the words encoded in the input. To predict each value in the output layer *(y)*, each weight vector of every neuron in that layer (V_{w0}) is multiplied by the vector generated by the hidden layer (V_{wi}), yielding the output values.

c. *Computation for Output Neurons*: As the above computation yields real values, they must be transformed into probability values to determine the "best" output word. To achieve this, a SoftMax-like function is applied to the previously computed value (*xi*). This function is a generalization of the logistic (or *exponential*) function, commonly employed to represent categorical functions.

$$\text{Soft Max}(x_i) = \frac{e^{x_i}}{\sum_{j=1}^{n} e^{x_j}}$$

In essence, it generates a probability distribution encompassing a set of potential outcomes, guaranteeing that each value falls within the range of 0 to 1.

To elaborate, the SoftMax function calculates the likelihood of an output word (w_o) given an input word (w_i). It does this by taking the exponential of the output neuron's result and dividing it by the sum of the exponentials of all the output values (represented by V vectors). In other words:

$$P(w_o \,|\, w_i) = \frac{e^{V_{wo}V_{wi}^T}}{\sum_{w=1}^{W} e^{V_w V_{wi}^T}}$$

After generating the prediction for the optimal output word (denoted as "w_o" in the preceding stage), each prediction is then compared to the vector representing the true input encoding. Typically, this input encoding is a *one-hot* vector consisting of zeros with only one non-zero value corresponding to the position of the word within the vocabulary. The error is subsequently calculated as the difference between the output probability and the input vector. Network weights are updated using gradient-based *back-propagation* methods (Sherstinsky, 2020).

Following this, within a given context, the word prediction process unfolds in three sequential steps:

1. The system searches for the *embedding* of the target word.

2. It computes the prediction for that word.

3. Finally, the word is projected into the output vocabulary.

It's important to note that the number of *embedding dimensions* or *features* corresponds to "N," which is also the number of neurons in the hidden layer.

One of the challenges associated with *embedding* models of this kind is their limited ability to handle sequences and capture the contextual nuances in which words are used. This limitation can potentially lead to changes in the *embeddings*, altering their intended meaning. Nonetheless, it's worth highlighting that certain fundamental concepts from such models, such as simple *feed-forward* networks and SoftMax functions, serve as foundational building blocks for more sophisticated methods.

2.7 RECURRENT NEURAL NETWORKS

When you read a sentence, your comprehension doesn't reset with each new word. Instead, you retain information from previous words to grasp the meaning of the word you're currently reading. This is why more conventional word *embedding* learning models like *Word2Vec* fall short—they struggle to encapsulate the rich contextual information that emerges from sequences of words within a text.

In tasks like language translation, where sentences consist of sequences of words, the current observation relies on preceding observations. Consequently, these observations are interdependent, and their independence cannot be assumed, as traditional machine learning methods, including neural networks, do. These conventional approaches lack the capability to retain historical information or recall past events, essentially having no memory of prior occurrences. To address this limitation, RNNs emerged, introducing the concept of *memory* within neural networks by establishing dependencies between data points. As a result, RNNs (Graves, 2012) can be trained to capture context-based concepts and learn recurring patterns (Sherstinsky, 2020).

2.7.1 Simple Recurrent Neural Networks

RNNs achieve memory through a feedback loop within their cells, marking a fundamental departure from traditional neural networks. This feedback loop facilitates the transmission of information within a layer, a key distinction from *feed-forward* networks, where information exclusively flows between layers.

In Figure 2.6, a portion of the RNN denoted as A takes in an input, x_t, and generates an associated value, h_t. The presence of a cyclic structure facilitates the seamless transfer of information between different time steps within the network.

RNNs are tasked with the challenge of determining which information is sufficiently relevant to retain in memory. In order to accomplish this, various types of RNNs have emerged, including the conventional RNN and the LSTM network. For instance, Figure 2.7 illustrates a generic network designed for translating from Spanish to English.

With each passing moment, the model receives a new word from the sentence that needs translation into English. This iterative process continues for the remaining words, making the translation task inherently recursive. Additionally, the model retains a memory of the previously translated word to establish context for subsequent translations.

One notable departure from the conventional representation is that the model is no longer depicted from left to right; instead, it is illustrated from bottom to top. Why is this change made? We reserve the horizontal axis to signify the passage of time. In our context, a time step represents the unit of action or time during which the model performs a

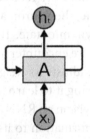

FIGURE 2.6   A RNN with cycles.

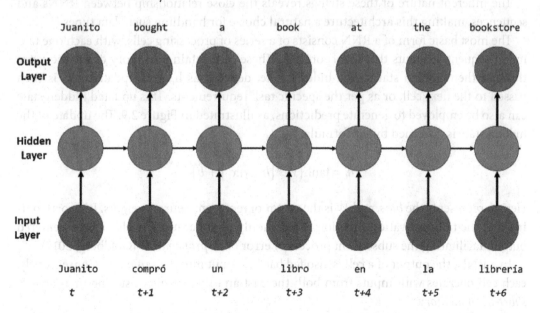

FIGURE 2.7   A RNN to translate words from a Spanish text.

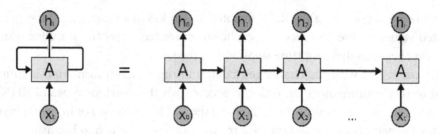

FIGURE 2.8   An unfolded RNN.

specific task. In our case, each time step corresponds to one word. Consequently, when translating a sentence consisting of five words, we will encounter five distinct time steps.

We begin by examining the initial layer, known as the *input layer*, connecting to a hidden layer marked in green. In contrast to an FNN, where the neurons in the hidden layer operate independently, in a RNN, the neuron from the previous time step is linked to the subsequent neuron in the same layer (as indicated by the arrow). Consequently, as we move forward in predictions, each neuron not only receives the current English word but also retains context information. Over time, the neuron accumulates and utilizes this context, evolving from mere memory into a dynamic state. Therefore, we refer to this transformed information as the "state." Eventually, we encounter an output layer responsible for executing the translation task. The network adjusts its weights based on prediction errors using the *back-propagation* algorithm, allowing it to learn and improve over time.

To gain a deeper understanding, envision an RNN as a series of interconnected instances of the same network, each passing information to its successor. Let's consider the consequences of unfolding this cyclic structure within the RNN, as depicted in Figure 2.8.

The inherent nature of these strings reveals the close relationship between RNNs and sequences, making this architecture a natural choice for handling such data types.

The most basic form of a RNN consists of a series of processing cells, with each one taking in sequential inputs that incorporate x_t. These cells retain a memory of the sequence through their hidden states. The hidden state, denoted as h_t, gets updated and is then passed to the next cell, or as per the specific task requirements. This updated hidden state can also be employed to generate predictions, as illustrated in Figure 2.9. The update of the hidden state is governed by the formula:

$$h_t = \tanh\left(W * [h_{t-1}, x_t], + b\right)$$

Here, b represents the *bias*, and W is the vector of recurrent neuron weights. It's worth noting that the tanh activation function is commonly used because it yields a zero-centered output, facilitating the subsequent process of error *back-propagation* (Goldberg, 2017).

In a RNN, the output of a cell is also fed back as input into the same cell. Consequently, each cell operates with inputs from both the past and the present, resulting in a form of *short-term memory*.

For a clearer understanding, let's delve into the feedback loop of an RNN cell, as depicted in Figure 2.10. The length of the *unrolled* cell corresponds to the number of

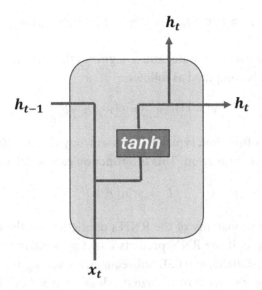

FIGURE 2.9   Operation of a RNN cell.

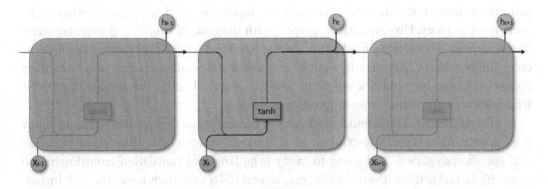

FIGURE 2.10   Input transfer to each cell of the RNN.

time steps in the input sequence. For instance, if the sequence comprises three words, the network unfolds into three interconnected subnetworks. Conceptually, this represents the network's *memory*, capturing information from prior steps. In contrast, traditional neural networks employ distinct parameters at each layer, whereas an RNN employs the same parameters across all layers. This parameter sharing indicates that the network is performing the same task with different inputs, thereby reducing the total number of learnable parameters.

We can observe how past observations traverse through the deployed network as a hidden state. In each cell, we blend the input from the current time step, denoted as x (current value), with the hidden state h from the previous time step (past value), and a bias term b. We combine these with the weights of the recurrent neuron (W_h) and the input neuron (W_x). Subsequently, these values are subjected to an activation function, which determines the hidden state for the current time step. Therefore, for instance, employing an activation function f (such as Sigmoid, TANH, or ReLU), the current state is computed as follows:

$$h_{t+1} = f(x_t, h_t, w_x, w_h, b_h) = f(w_x x_t + w_h h_t + b_h)$$

Here, h_{t-1} represents the previous state, and x_t signifies the current input. The output y at time t, with weight W_y, is computed as follows:

$$y_t = f(h_t, w_y) = f(w_y \cdot h_t + b_y)$$

To train an RNN, a loss function, typically cross-entropy (Baron, 2019), is essential, often accompanied by a SoftMax function. This *loss* function can be calculated as follows:

$$L = -\ln(p_c)$$

where p_c represents the probability of the RNN's prediction for the correct class (positive or negative). For example, if the RNN predicts a text as positive with a 95% probability, the loss would be: $L = -\ln(0.95) = 0.051$. Subsequently, training the RNN involves utilizing gradient descent algorithms to minimize this loss, such as Back-Propagation Through Time (Ekman, 2022).

One of the advantages of an RNN lies in its ability to process sequential data, discern patterns in historical data, and accommodate inputs of varying lengths, owing to its short-term memory. However, RNNs grapple with the issue of *vanishing gradient descent*. This occurs when gradients, used to update the weights during *back-propagation*, become exceedingly small. Consequently, multiplying weights by gradients close to zero hinders the network from learning new weight patterns. In practical terms, this implies that RNNs tend to forget information from longer data sequences.

In a broad context, RNNs find applications in various prediction scenarios: one-to-many (e.g., generating a textual description of an image), many-to-one (like classifying text into specific categories), and many-to-many (e.g., language translation from English to Spanish). To tackle these diverse challenges, several RNN variations have emerged, including Bidirectional Recurrent Neural Networks, Gated Recurrent Units, and LSTM networks.

2.7.2 Long Short-Term Memory Networks

An LSTM network, often referred to as LSTM, stands out as a specialized type of RNN designed to address a critical issue plaguing simple RNNs: the vanishing gradient problem. This problem arises due to the loss of information over extended periods, a consequence of long-term dependencies (Bohnet et al., 2018).

LSTMs exhibit a chain-like structure, but they differ from basic RNNs by employing four distinct layers that interact in a unique manner, as depicted in Figure 2.11.

At the heart of the LSTM lies the *cell state* (c), which traverses the entire chain, facilitating the seamless flow of information with minimal linear transformations through three crucial gates. Consequently, the cell state serves as the repository for the network's long-term memory.

Information flows through three *gates*: the forget gate, the input gate, and the output gate, illustrated in Figure 2.12. Each of these gates consists of a *sigmoid* layer paired with an element-wise multiplication operation (Liu and Perez, 2017).

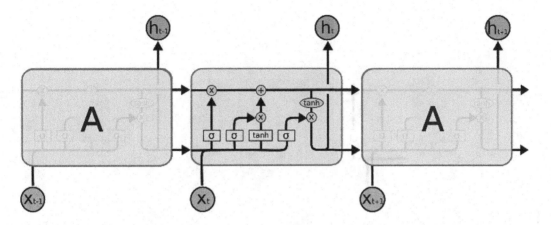

FIGURE 2.11   A LSTM module with four interaction layers.

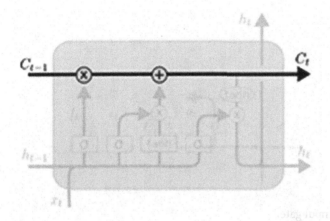

FIGURE 2.12   Information flow in a LSTM cell.

The *sigmoid* layer serves a crucial role in generating values between 0 and 1, effectively regulating the passage of information. A value of 0 signifies "do not allow through," while a value of 1 denotes "let everything pass." In this manner, these gates function as filters, controlling the flow of data and determining what information to retain or discard:

- **Forget gate**: The forget gate determines which portion of the long-term memory should be preserved. It employs a sigmoid function to assess the significance of the cell state, C_{t-1}. The resulting output ranges from 0 to 1, indicating the extent of information retention. A value of 0 signifies no information retention, while 1 implies the retention of all cell state information (Figure 2.13).

    The output is determined by combining the current input x, the hidden state h from the previous time instant, a bias b, and the respective weights W:

$$f_t = \sigma(W_{f,x}x_t + W_{f,h}h_{t-1} + b_f)$$

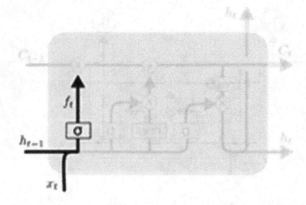

FIGURE 2.13   A forget gate.

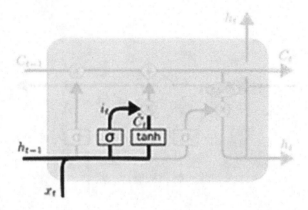

FIGURE 2.14   An input gate.

- **Input gate**: The input gate decides which information should be incorporated into the cell state, thereby influencing long-term memory. It employs a *sigmoid* layer to determine which values are eligible for updates (Figure 2.14).

$$i_t = \sigma\left(W_{i,x}\,x_t + W_{i,h}\,h_{t-1} + b_i\right)$$

Now we must update the previous cell state, C_{t-1}, in the new cell state, C_t. To do this, we multiply the previous state by f_t, forgetting the things we decided to forget before, and then adding $i_{t+}C_t$ to it.

$$C_t = \tanh\left(W_c * [h_{t-1}, x_t] + b_C\right)$$

These newly computed candidate values are scaled based on our earlier decisions regarding the importance of each state value. In the context of a language model, this is where we decide which information from previous words to retain and which to discard, as determined in the preceding steps (Figure 2.15).

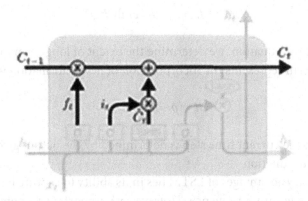

FIGURE 2.15 The flow from the input gate to the output gate.

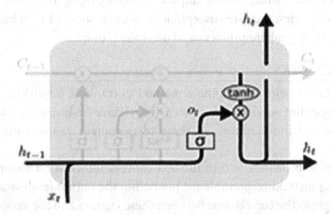

FIGURE 2.16 Output gate.

- **Output gate**: This component plays a crucial role in determining which aspects of the cell state contribute to the final output. It is responsible for managing short-term memory based on the cell state, albeit in a filtered manner. To achieve this, we employ a sigmoid layer that determines which portions of the cell state we want to incorporate. Subsequently, the cell state is passed through a hyperbolic tangent (tanh) function to confine its values within the range of –1 to 1. This transformed state is then multiplied by the output of the sigmoid gate, allowing us to selectively generate the desired components (Figure 2.16).

  Following this, a tanh layer generates a vector containing new candidate values (Ct) that could potentially be added to the cell state. These components are combined to produce an updated cell state:

$$o_t = \sigma(W_{o,x}x_t + W_{o,h}h_{t-1} + b_o)$$

As depicted, the three gates are represented by the same underlying function, with only variations in weights and biases. The cell state is updated by the interplay of the forgetting gate and the input gate:

$$c_t = (f_t \circ c_{t-1} + i_t \circ \tanh(h_{t-1}))$$

In the first term of the equation, we determine the extent of long-term memory retention, while the second term represents the incorporation of new information into the cell state:

$$h_t = o_t \circ \tanh(c_t)$$

The hidden state at the current time step is determined by the output gate's output in conjunction with a tanh function.

One of the primary advantages of LSTM lies in its ability to capture both long-term and short-term patterns within a sequence. However, it's important to note that LSTMs are computationally more intensive than traditional RNNs, which can result in longer training times. Additionally, since LSTMs employ the *Back-Propagation* Through Time algorithm for weight updates, they are susceptible to issues associated with back-propagation, such as the vanishing gradient problem and dead *ReLU* units.

2.8 AUTOENCODERS

Imagine you're tasked with developing a method to translate English text into Chinese. Creating this algorithm would require extensive reliance on historical English texts that have been translated into Chinese, as these texts contain valuable links between the languages.

Each observation or *feature* within the text conveys information about the input. The machine learning method responsible for predicting the output (each word in the translated text) must grasp the correlations between these *features* and the various words in the output text.

However, many machine learning algorithms encounter issues when faced with noisy or inconsistent data. This problem arises due to their limited depth of understanding of the data.

One potential solution to this challenge lies in creating a more abstract representation of the data. For many tasks, it's often impossible to determine which *features* should be extracted. To tackle this, we can present the data to the computer and let it autonomously learn the representation. This task is known as representation learning or RL (Achille and Soatto, 2017). This process transforms high-dimensional data, such as input texts, into low-dimensional representations, offering several advantages:

- It simplifies the detection of patterns and anomalies.

- It enhances our comprehension of overall data behavior.

- It reduces data complexity to filter out noise, making it particularly valuable for supervised machine learning techniques.

In generating these representations, RL focuses on the properties learned by the layers of a neural network, often referred to as "activations." Importantly, these representations remain independent of the specific optimization process used. To achieve a strong representation, it's typically crucial to eliminate two factors prevalent in data distributions:

1. **Variance**: This can be likened to the sensitivity that could potentially result in dramatic output variations. Any model we construct must exhibit resilience to variance.

2. **Entanglement**: This refers to how one *embedding* connects or correlates with other *embeddings*. Such intricate connections make the data highly complex and challenging to decipher. To mitigate this, we should pinpoint variables where relationships are simpler to understand.

If none of the conventional architectures explicitly enforce invariance and incentivize disentanglement (Achille and Soatto, 2017), how can these crucial properties spontaneously emerge within deep learning networks trained through simple optimization?

To address this question, we must tackle two fundamental issues:

1. Utilizing information theory, we can demonstrate that achieving invariance in deep neural networks, often referred to as "many-layered" networks, is equivalent to cultivating a minimal representation within the computational process. This can be accomplished by stacking layers in the network and introducing controlled noise into the computation.

2. Employing empirical loss information decomposition, we can establish that mitigating overfitting, a common challenge in deep learning, can be achieved by constraining the information content stored within the neural network's weights.

2.8.1 The Information Bottleneck

The concept of an information bottleneck compels the extraction of relevant information by *compressing* the volume of information that can traverse the entire network. This necessitates the learning of a compressed representation of the input data.

This compression not only reduces the dimensionality of the data but also simplifies its complexity. Consequently, a neural network can eliminate irrelevant details from noisy input data, akin to squeezing information through a bottleneck, leaving only the *features* most pertinent to general concepts intact.

2.8.2 Latent Variables

A *latent variable* is a random variable that remains hidden from direct observation but plays a pivotal role in determining the distribution of data. These latent variables offer us a foundational, low-level representation of high-dimensional data, essentially providing an abstract depiction of data distribution.

Now, why do we need latent variables?

All machine learning methods must solve the problem of learning probability distributions $p(x)$. These distributions are restricted to a limited set of high-dimensional data (*dataset*) generated from them.

For instance, consider the task of learning the probability distribution of texts for translation. It entails defining a distribution capable of capturing intricate correlations among the words composing each text. Directly modeling this distribution is a formidable and laborious task, even with virtually limitless time at hand. Here's where latent (unobserved) variables come into play. We can introduce a latent variable z and define a conditional distribution, $p(x|z)$, for the data. In the context of text translation, z could encapsulate hidden representations of word *features*. In simpler terms, models that employ these latent variables enable a generative process that mirrors the data generation process. This is commonly referred to as a *generative model*.

This implies that when we aim to generate a new data point, we first obtain a sample $z \sim p(z)$ and subsequently use it to draw a new observation x from the conditional distribution $p(x|z)$. Simultaneously, we can assess whether the model effectively approximates the data distribution $p(x)$. On the other hand, mathematical models featuring latent variables are, by their very definition, latent variable models. These latent variables possess significantly lower dimensions than the observed input vectors, resulting in a compressed representation of the data.

In reinforcement learning (RL), these principles find application in various methods, including *Multi-Layer Perceptrons*, CNN, and *Autoencoders* (Ekman, 2022). However, *autoencoders* gain paramount importance in contemporary language models, serving as the foundational concept for the *transformers* architectures discussed later (Rothman, 2022).

Autoencoders, which are unsupervised learning methods rooted in neural networks, can be effectively trained for reinforcement learning tasks. They achieve this by employing a combination of an *encoder* and a *decoder*, as depicted in Figure 2.17. Typically, the training process for *autoencoders* involves learning algorithms that compare the network activation of the input with the activation of the reconstructed input.

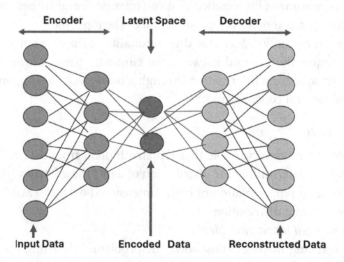

FIGURE 2.17 *Autoencoder* architecture.

While *autoencoders* are commonly used for tasks such as dimensionality reduction and *feature* learning, they also possess the capability to construct generative models. These models have the unique ability to generate new data points. This is accomplished by compressing the information into an *"information bottleneck,"* where only the most relevant *features* are extracted from the entire *dataset*. These extracted representations can then be utilized to generate fresh data.

In simpler terms, an *encoder* serves as a function responsible for reducing the input into distinct representations. Subsequently, a *decoder*, also a function, takes these learned representations from the *encoder* and transforms them back into the original format.

2.8.3 *Autoencoder* Architecture

Now, let's delve into the architecture of an *autoencoder*, which despite having various designs, shares three fundamental components, as illustrated in Figure 2.17:

1. **Encoder**: This component is responsible for compressing the input information. It achieves this by stacking layers of neurons, reducing the number of neurons used in each layer.

2. **Latent Space** (or "Information Bottleneck"): This represents the minimal space within the neural network where the information is encoded, essentially serving as a compressed space).

3. **Decoder**: The *decoder*'s role is to allow the network to decompress the representations and reconstruct the data from its encoded form. The output is then compared with the true values.

An ideal *autoencoder* model must strike a delicate balance:

1. *It should be sensitive enough to the inputs to generate accurate reconstructions.*

2. *It should be insensitive enough to inputs that the model does not intend to memorize or overfit from during the training.*

This equilibrium compels the model to retain solely the essential data variations required for input reconstruction, avoiding the entanglement of redundancies within the input. In most instances, this process entails formulating a loss function that consists of two key components: one term encourages the model's responsiveness to inputs [i.e., the reconstruction loss, denoted as $\mathcal{L}(x,\hat{x})$], and a second term discourages memorization or *overfitting* (typically achieved through the inclusion of a *regularizer*).

$$\mathcal{L}(x,\hat{x}) + regularizer$$

Typically, a scaling parameter is introduced, allowing for the *fine-tuning* of the trade-off between these two objectives.

2.8.4 Types of *Autoencoders*

There are different structures that *autoencoders* can adopt. The most successful ones include the following:

a. **Incomplete *Autoencoder*:**

This model limits the number of nodes present in the hidden layers of the network, which restricts the amount of information that can traverse the network. By penalizing the network according to the reconstruction error, the model can learn the most important *features* of the input data and how to best reconstruct the original input from an encoded state. Ideally, this encoding will learn and describe the latent *features* of the input data (Figure 2.18).

Since neural networks are capable of learning nonlinear relationships, this can be seen as a more powerful nonlinear generalization of principal component analysis (Baron, 2019). While principal component analysis attempts to discover a hyperplane in low dimensions that describes the original data, an *autoencoder* is able to learn a complex nonlinear surface that can be used to describe observations in a low-dimensional space and decoded correspondingly in the original input space.

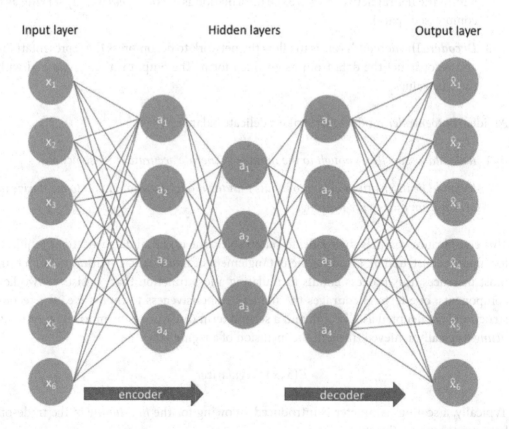

FIGURE 2.18   Architecture of an incomplete *Autoencoder*.

b. **Sparse *Autoencoder*:**

This model limits the number of neurons in the hidden layer that are activated to force the model to learn more complex patterns. This is achieved through the sparsity constraint, where a term is added to the loss function that forces the *autoencoder* to decrease the number of active neurons during training.

For any given observation, the network is incentivized to learn an *encoder* and a *decoder* that only rely on a few neurons. It is important to note that the individual nodes of a trained model that are activated are data dependent, so different inputs will produce activations of different nodes throughout the network (Figure 2.19).

A consequence of this phenomenon is that the network becomes sensitive to specific *features* of the input data through its hidden layer nodes. Unlike an incomplete *autoencoder*, which employs the entire network for each observation, a *sparse autoencoder* is compelled to selectively activate certain network regions based on the input data. Consequently, this limitation curtails the network's capacity to memorize input data while preserving its ability to extract data *features*.

This distinction empowers us to address latent space representation and network regularization as separate entities. We can opt for a particular latent space

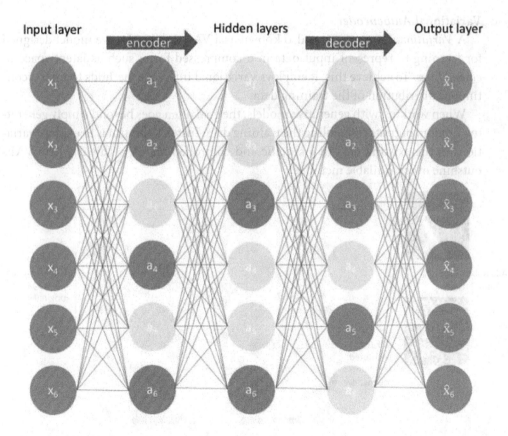

FIGURE 2.19   Architecture of a sparse *Autoencoder*.

representation (e.g., *embeddings*) that aligns with the data context while enforcing regularization through a sparsity constraint.

There are two approaches to impose this sparsity constraint, involving the measurement of hidden layer activations for each training *batch* and incorporating a loss function to penalize excessive activations: L1 regularization and KL (*Kullback-Leibler*) divergence (Ekman, 2022).

c. **Denoising *Autoencoder*:**

This model introduces subtle data *corruption*, where the uncorrupted data serves as the target output. This compels the network to identify intricate patterns by introducing noise from a random Gaussian distribution, thus modifying the input data to emphasize pertinent data patterns (Figure 2.20).

Through this approach, the model learns a vector field that transforms the input data into a concentrated, low-dimensional region. If this region accurately represents the natural data, the model effectively mitigates the added noise.

It's essential to acknowledge that this vector field predominantly performs well within regions observed during training. In regions far from the natural data distribution, the reconstruction error can be substantial and may not align with the true distribution.

d. **Variational *Autoencoder*:**

A *Variational Autoencoder*, also known as a VAE, is a generative model designed for learning to represent input data in a compressed form, such as latent space or *embeddings*. To achieve this, it employs variational inference methods to infer a continuous distribution of the training data.

When working with generative models, the goal often goes beyond simply generating random outputs resembling the training data. Instead, the aim is to explore variations in the existing data in a specific and controlled manner. This is where VAEs outshine other available methods.

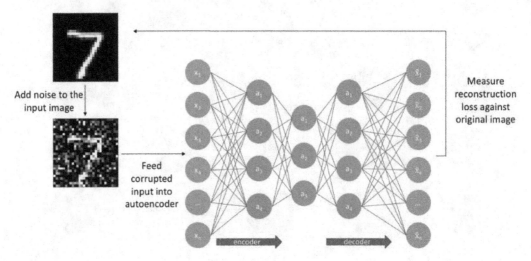

FIGURE 2.20 Architecture of a denoising *Autoencoder*.

VAEs possess a distinctive and crucial characteristic: their latent spaces are inherently continuous, enabling effortless random sampling and interpolation. This is accomplished by the *encoder* producing two vectors of size "*n*": a vector of means represented as "μ" and a vector of standard deviations represented as "σ." These parameters form a Gaussian distribution, from which the *decoder* can draw samples to create the latent space. Consequently, the *decoder* can generate synthetic data that closely resembles the actual data.

These parameters represent a vector of random variables of length "*n*," where the "i-th" element of "μ" and "σ" corresponds to the mean and standard deviation of the "i-th" variable, "X_i." Consequently, stochastic sampling is employed to obtain the sample coding, which is subsequently passed to the *decoder*.

This stochastic generation implies that even for the same input, with identical mean and standard deviation values, the actual encoding will exhibit slight variations on each pass due to the sampling process. Essentially, the vector of means dictates the central point where the encoding of an input should be positioned, while the standard deviation controls the extent to which the encoding can deviate from the mean. Since the encodings are randomly generated from anywhere within the distribution, the *decoder* learns that not only does a single point in latent space correspond to a sample of that class, but all nearby points also relate to it. This empowers the *decoder* to decode not only unique and specific encodings in the latent space but also those that exhibit subtle variations, as the *decoder* is exposed to a spectrum of encoding variations for the same input during training.

To ensure that *embeddings* are as close to each other as possible for similar words, the loss function introduces KL divergence. KL divergence quantifies the dissimilarity between two probability distributions, with minimizing it implying that the parameters of the probability distribution (μ and σ) should resemble the target distribution:

$$\sum_{i=1}^{n} \sigma_i^2 + \mu_i^2 - \log(\sigma_i) - 1$$

For a VAE, the KL loss equals the sum of all KL divergences between the components $X_i \sim N(\mu i, \sigma i^2)$ in "*X*" and the standard normal distribution. As evident, this is minimized when "$\mu i = 0$" and "$\sigma i = 1$."

In essence, this loss incentivizes the *encoder* to evenly distribute all encodings around the center of the latent space, penalizing any attempts to *group* them separately in specific regions away from the origin

2.9 GENERATIVE ADVERSARIAL NETWORKS

While an *autoencoder* excels in data compression and precise reconstruction of original inputs, there are numerous scenarios and applications that require the generation of realistic data, such as images, that are indistinguishable from real samples. Achieving this realism increases diversity and thereby augments the available training data.

One approach to tackle such a generative task involves leveraging a model known as a *Generative Adversarial Network* (GAN). GANs belong to the formidable realm of neural networks and are employed for generative modeling through unsupervised learning, as highlighted by key references such as Bengio (2014) and Babcock and Bali (2021). At its core, a GAN consists of a dynamic interplay between two neural network models engaged in a competitive dance, capable of comprehending, capturing, and replicating variations within a *dataset*. These two networks assume the roles of the *generator* and the *discriminator*, each with its own unique function. The generator, typically implemented as a CNN, takes random noise as input and crafts an image that aligns with the desired domain. In contrast, the *discriminator's* role is to discern whether the generated data, such as an image, is authentic or not. In other words, it evaluates whether the output genuinely belongs to the intended distribution or not.

This framework not only facilitates the generation of synthetic data but also enhances the performance of learning models as a data augmentation technique, reducing generalization errors. It achieves this by generating new, artificial yet plausible examples from the input problem domain on which the model is trained.

In theory, GANs operate within a framework rooted in game theory. Within this framework, a generating neural network engages in a strategic competition with an adversarial network. The role of the generator network is to produce samples directly, while its counterpart, the discriminative network, strives to distinguish between samples originating from the training data and those generated by the generator (as depicted in Figure 2.21).

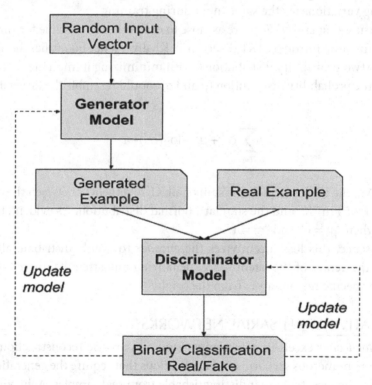

FIGURE 2.21 Composition of a GAN.

In contrast to *autoencoders*, GANs exhibit the remarkable ability to generate high-quality, realistic content. GANs leverage their adversarial architecture, enabling the generator network to learn and produce synthetic samples that closely resemble the patterns found in the original training data. This inherent capability makes GANs well-suited for a wide array of tasks, including text and image generation, among others. Unlike *autoencoders*, which often generate results closely aligned with the training data, GANs have the unique capacity to learn and generate samples that faithfully capture the underlying distribution of the original *dataset*.

2.9.1 The Generative Model

In the realm of generative models, a pivotal role is played by the generative model itself. This model takes as its input a random vector of fixed length, a vector drawn randomly from a Gaussian distribution, and then works its magic to generate a sample within the defined domain (as illustrated in Figure 2.22). It's important to note that this multidimensional vector space, arising from the generative process, eventually houses points that correspond to the problem domain. In essence, this space serves as a compressed representation of the data distribution.

Within this vector space lie latent (hidden) variables, which carry significance for the domain but remain elusive to direct observation.

2.9.2 The Discriminative Model

In contrast, the discriminative model takes an example from the domain, whether it's real or generated, and plays a pivotal role in predicting a binary class label: either real or fake (as depicted in Figure 2.23). Real examples are drawn from the training *dataset*, while the generated instances are the handiwork of the generator model. The discriminator functions as a conventional and well-understood classification model.

Following the training phase, the discriminator model's role concludes, as our primary interest lies in the generator model. It's worth noting that the generator, having honed its ability to extract *features* effectively from domain examples, can sometimes find further utility. In particular, its *feature* extraction layers can be employed in transfer learning applications that involve similar or identical input data.

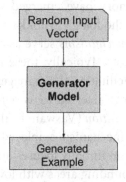

FIGURE 2.22   The generative component of a GAN.

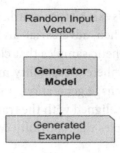

FIGURE 2.23 The discriminative component of a GAN.

2.10 ATTENTION MODELS

Generative models, such as *autoencoders* or GANs, have proven to be highly efficient in generating synthetic data from real samples. However, they often lack the ability to *focus* on the most relevant *features* when processing input samples or generating output. In simpler terms, these approaches tend to *overlook* specific details, which can significantly impact the quality of training and the generated samples. To address this issue, incorporating mechanisms that allow these networks to concentrate on the pertinent aspects of the data becomes essential.

Have you ever wondered how you can have a meaningful conversation with a friend at a noisy party, amidst a cacophony of voices? This scenario presents what is known as the *"cocktail party problem."* Most of our cognitive processes naturally prioritize attention to a singular activity at any given moment (Bermúdez, 2020). This ability to direct our focus toward specific information while filtering out distractions extends to various complex cognitive tasks, such as deciphering crucial words in a text for translation or identifying the most critical elements in an image.

The core cognitive processes that help us solve the "cocktail party problem" are *attention and short-term memory*. Being able to decode information by paying attention to fragments rather than processing the entire data stream in one go enables us to respond in real-time.

In the realm of computational systems, challenges abound in processing and storage. Computers operate in a binary world of 0s and 1s, yet the concept of *attention*, as understood by humans, can be incorporated into them through computational techniques. Consequently, attention and memory have emerged as pivotal components in recent advancements in deep learning methodologies (Bokka et al., 2019).

Within this context, *attention mechanisms* serve as computational tools that enhance the relevance of specific components. Typically, these mechanisms zero in on elements within a network's architecture, facilitating the management and quantification of interdependencies among input elements, referred to as *self-attention*, or between input and output elements, termed *general attention* (Vaswani et al., 2017).

This mirrors the visual attention mechanism employed by the human brain. For instance, when we observe an image, our brain initially focuses on a specific region with high-resolution detail while perceiving surrounding areas with lower resolution. As our brain comprehends the image, it dynamically adjusts the focal point to encompass all aspects fully.

Attentional models assess inputs to identify the most crucial components and assign them weighted importance, often referred to as *attentional weights*. For instance, in the context of translating a sentence from one language to another, the model highlights and assigns higher weights to the most vital words, enhancing the accuracy of the output prediction.

Initially designed to enhance machine vision and *autoencoder*-based neural machine translation techniques, attention models have significantly contributed to the field of NLP. They have facilitated the creation of fixed-length vector representations, improving the performance of various tasks such as translation and comprehension. This integration of attention mechanisms in NLP has catalyzed breakthroughs, giving rise to *Transformer* architectures and subsequently, LLMs like BERT and GPT, among others.

Traditionally, NLP applications employing neural models relied on *encoder/decoder* architectures, often based on RNN or LSTM. However, a notable drawback of this approach is the challenge posed by long-range dependencies. When the *encoder* attempts to summarize longer sentences, it often produces subpar summaries, leading to lower translation quality. This problem is called *long-range dependence* of an RNN/LSTM.

Consider, for instance, the task of predicting the next word in a sentence where the context lies a few words behind: "*Despite being originally from Italy, since he was raised in Croatia, he feels more comfortable speaking Croatian.*" In such cases, when predicting the word "Croatian," the words "raised" and "Croatia" should carry more weight, while "Italy," although the name of another country, should be less influential. Is there a way to preserve all relevant information in input sentences while creating the context vector?

Thus, at any instant that the model generates a sentence, it searches for a set of positions in the hidden states of the *encoder* where the most relevant information is available. This mechanism is called "attention" in *deep learning* (Deng and Liu, 2018).

2.10.1 *Encoder-Decoder* Paradigm

It's worth noting that the *autoencoders* we've encountered so far are essentially specialized instances of *encoder-decoder* models, where both input and output mirror each other. These models belong to a family that excels at transforming data points from one domain to another via a two-stage network. The *encoder*, embodied by the encoding function $z = f(x)$, efficiently compresses the input into a latent space representation. Subsequently, the *decoder*, represented as $y = g(z)$, uses this representation to predict the output

The *encoder-decoder* architecture is typically constructed using RNNs and finds popularity in diverse applications such as machine translation and sequence-to-sequence prediction, often referred to as *Seq2Seq*. An advantage of this architecture lies in its ability to separate the *encoder* from the *decoder*, accommodating varying sequence lengths.

Models with varying sequence lengths find applications in tasks like sentiment analysis, where a sequence of words yields a numeric output, or in image *caption* generation, where an image input generates a sequence of words.

In Figure 2.24, we observe an *encoder-decoder* model designed for machine translation. The *encoder*, depicted as a blue rectangle, consists of an input layer and an LSTM. This *encoder* receives a Spanish sentence and produces an *embedding* vector, which encapsulates

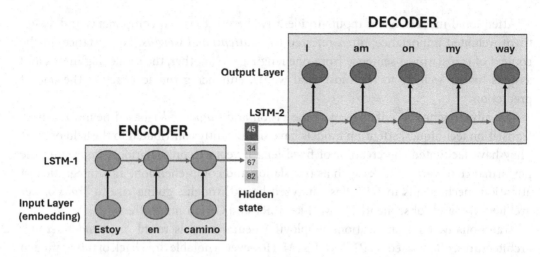

FIGURE 2.24   An illustration of an *encoder-decoder* model for translation.

the essence of the complete sentence in its hidden state at the final LSTM time step. The *decoder*, in turn, takes this hidden state as input and generates the English translation in the form of a sequence of words.

However, one limitation of this architecture is evident as sentence length increases. The model struggles to capture all the nuances, especially in longer sentences. It tends to forget portions of the sentence, mainly because it relies solely on the representation at the end of the *encoder*. But different segments of the input sequence might hold more relevance at different stages of generating the output sequence. This is where the concept of *attention* comes into play.

2.10.2 Attention to Sequence Models

Consider the scenario where you need to translate an English text sequence into French. Figure 2.25 illustrates a heatmap that showcases where a model directs its attention during sentence translation.

The horizontal axis represents the input English sentence, while the vertical axis represents the generated French translation. Brighter areas signify higher attention. It's noteworthy that sometimes a translated word *attracts* attention from several English words.

For instance, to generate the word "*accord*," the model primarily focuses on "*agreement*." The power of attention becomes evident when the model translates an entire sentence like "European Economic Area" into "zone economique européenne." English employs adjectives before nouns, whereas French follows the opposite pattern. Hence, the model shifts its attention to words that appear after. This highlights how translation relies not just on individual words but also their contextual placement within the sentence.

In this context, attention transitions from the *encoder* to the *decoder*. The *decoder* generates translated words sequentially, with each output word influenced by all input words but with varying degrees of attention. The idea is to capture these varying *weights* of attention.

Take, for instance, the sentence "*a student in the class asked.*" Figure 2.26 reveals where the model concentrates its attention when predicting the next word.

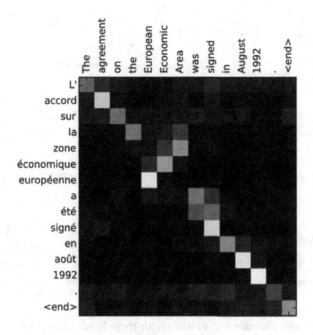

FIGURE 2.25   Attention focus in the translation of a sentence.

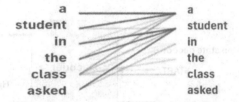

FIGURE 2.26   Attention in an input sentence.

The lines, when read from left to right, illustrate where a model directs its attention while *predicting* the next word in a sentence. The intensity of these lines signifies the *strength* or *weight* of this attention. For instance, when the model *anticipates* the next word after "*asked*," it places significant emphasis on "*student*." This emphasis is entirely logical, as identifying the entity or subject that is "asking" is pivotal to predicting what follows. In linguistic terms, the model zeroes in on the head of the noun phrase "a student in the class." In this manner, a plethora of such linguistic patterns can be effectively captured.

The critical question here is how the model determines where to focus its attention. This is determined through the calculation of an *alignment score*, which quantifies the attention each input word should receive. The shifting of attention profoundly affects the interpretation and, consequently, the results. Unlike traditional sequence models with fixed-length context vectors that struggle with longer input sequences, attention mechanisms provide a solution.

Consider the sentence "Where is Wally?" needing translation into Italian as "Dove è Wally?" Figure 2.27 demonstrates how the *encoder* processes words incrementally, producing multiple hidden states.

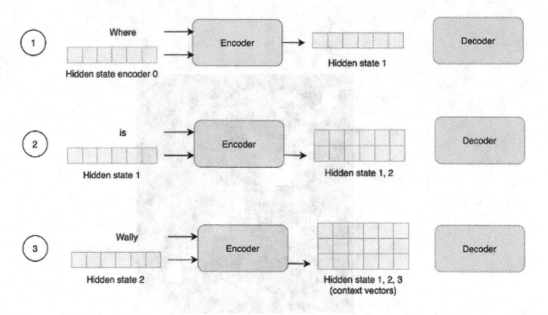

FIGURE 2.27   An *encoder-decoder* architecture with three hidden states for the "Where's Wally?" sequence.

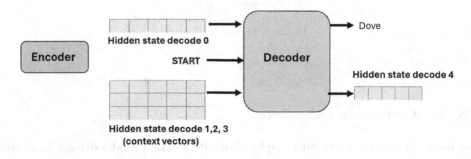

FIGURE 2.28   A *decoder* that produces the first output token, considering the hidden states of the *encoder*.

With an attention mechanism, all these hidden states are conveyed to the *decoder*, as opposed to only the final one (Figure 2.28).

Now, how does the *decoder* leverage these varying quantities of hidden states, depending on the input sequence length? This is where *attention mechanisms work* (Vaswani et al, 2017).

Attention generates a fixed-length context vector, achieved through weighted summation of *encoder* hidden states. Each weight reflects the attention directed toward a specific context when processing a particular input word (Figure 2.29).

It's essential to note that the attention mechanism operates just once within the model, serving as the vital link between the *encoder* and *decoder*. It takes the matrix of *encoder* hidden states and determines where to *focus* using *alignment scores* or weights.

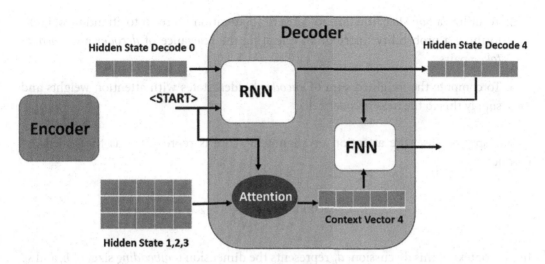

FIGURE 2.29   A *decoder* that generates the first token from context vectors.

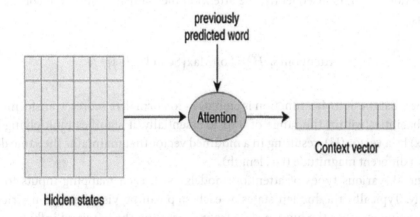

FIGURE 2.30   The attention mechanism from hidden states to context vector.

Essentially, attention *condenses* a list of hidden states into a single context vector, taking into account the current word being processed.

A simplified view of this process is depicted in Figure 2.30, where the *hidden states* and the previously predicted word contribute to the creation of a new *context vector*, guided by the attention focus.

In *Seq2Seq* models, each *decoder token* scrutinizes every *encoder* symbol to ascertain which input *tokens* demand more attention through weighted averages of their hidden states $H = [h_1 \, h_t \ldots h_l]^T$.

To integrate *attention* into *Seq2Seq* models, three steps are typically followed:

1. To calculate a scalar *attention score* (*weight*) for each pair of hidden *decoder* and *encoder* states (s_i, h_j) to gauge the relevance of *encoder token j* to *decoder token i*.

2. To utilize a *SoftMax* function to convert all attention scores into attention weights, forming a probability distribution reflecting the relevance of *decoder* and *encoder token* pairs.

3. To compute the weighted sum of *encoder* hidden states with attention weights and supply this to the next *decoder* cell.

In our specific case, the attention or alignment score is represented as the scaled dot product:

$$\text{Score}\left(s_i, h_j\right) = \frac{s_i^T h_j}{\sqrt{d_h}}$$

In the context of this discussion, d_h represents the dimension (*embedding* size) of h_j and s_i.

It's worth noting that in the numerator, we perform a dot product, effectively measuring the cosine similarity between s_i and h_j. Meanwhile, the denominator serves as a simple normalization factor. Consequently, the attention mechanism adjusts the context vectors as follows:

$$\text{Attention}\left(s_i, H\right) = \text{SoftMax}\left(\text{Score}\left(s_i, h_j\right)\right) * H$$

Remember that the SoftMax function is employed to normalize scores, transforming them into probabilities within the range of 0 to 1. Essentially, it signifies multiplying a scalar (SoftMax) by a vector (H), resulting in a modified vector that maintains the same direction but with a different magnitude (i.e., length).

In general, various types of attention models exist, each mapping inputs to outputs differently. Typically, the hidden states at each step can be visualized as matrices, with cells indicating whether the hidden state from processing the left word influences decoding the current word. This involves different components such as fonts, *encoders*, *decoders*, and *weights*. Common types of attention models include global store attention and self-attention.

2.10.2.1 Global Attention Model

This model aggregates input from all *encoder* and *decoder* states before evaluating the current state to generate the *context vector*, as depicted in Figure 2.31.

This approach uses every *encoder* step and previous *decoder* step to compute attention weights or alignment weights. Subsequently, each *encoder* step output is adjusted by these alignment weights to determine the context value, enabling the RNN cell to produce the *decoder* output.

It's important to emphasize that the alignment score is the crux of the attention mechanism as it quantifies how much attention the *decoder* should allocate to each *encoder* output when generating a new output.

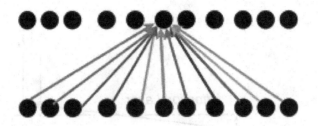

FIGURE 2.31   Global attention model.

The calculation of attention involves a weighted sum. A feed-forward network (FNN) calculates specific weights, considering the corresponding hidden state and the input word (*token*) being read by the model at that moment. This computation generally comprises three steps:

1. **Encoding**: The *encoder* encodes the input sequence. Each encoded time instant is assessed using the target decoding, and the scores are normalized using a SoftMax function. Typically, four possible evaluation functions can be used to produce a sequence of the same length (h_s).

2. **Decoding**: The *decoder* interprets the encoding, generating a target decoding (h_t).

3. **Alignment**: Each encoded time instant is assessed using the target decoding, and the scores are normalized using a SoftMax function. Typically, four possible evaluation functions can be used:

   a. *Dot product* between the target decoding and the source encoding.

   b. *Scalar product* between the target decoding and the weighted source coding.

   c. *Concatenation* of neural network processing of the source coding and the target decoding.

   d. *Softmax* of the weighted target decoding.

4. **Context vector**: The context vector is formed by applying the alignment weights to the source coding through a weighted sum.

5. **Final decoding**: the context vector and target decoding are concatenated, weighted, and transferred using a *tanh* function to obtain the final decoding.

The resulting encoding is passed through a SoftMax function to predict the probability of the next word in the sequence (y_t). Figure 2.32 illustrates the generation of alignment, information flow between the hidden and attention layers, and the context vector.

On the other hand, Figure 2.33 illustrates the operations involved in calculating global attention using a dot product evaluation function.

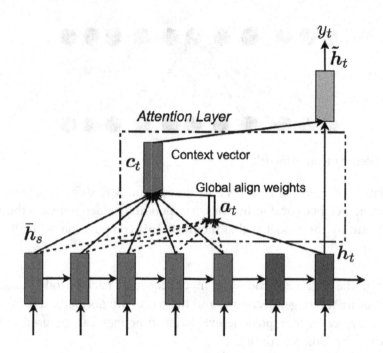

FIGURE 2.32   Alignment of the hidden layer with the attention layer to generate the context vector in global attention.

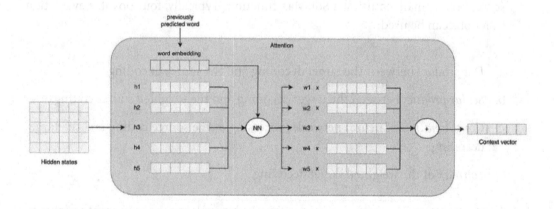

FIGURE 2.33   Global attention operation.

It's important to note that this model computes attention over the entire input sequence, hence its name "global attention." However, this approach can be computationally intensive and sometimes unnecessary.

2.10.2.2 Local Attention Model

In contrast to the global attention model, *local attention* only considers a subset of *encoder* positions when determining alignment weights. This significantly reduces computational complexity, as shown in Figure 2.34.

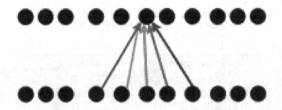

FIGURE 2.34   Local attention model.

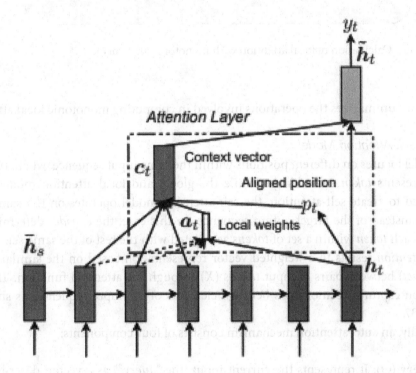

FIGURE 2.35   Alignment of the hidden layer with the care layer to generate the context vector in the local attention model.

The model calculates these weights and the context vector using the first aligned position and a selection of words from the *encoder*.

A typical example of a local attention model is depicted in Figure 2.35. It identifies a single aligned position (p_t) and utilizes a window of words from the source (*encoder*) along with the context vector the context vector (h_t) to compute alignment weights and the context vector. The store can typically be of two types:

- **Monotonic alignment**: It assumes that only specific information is relevant, setting the position (p_t) as t.

- **Predictive alignment**: It enables the model to predict the final alignment position, adjusting only the position (pt) as t is predicted by the model.

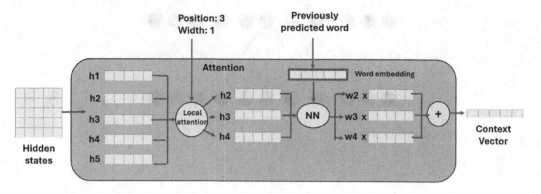

FIGURE 2.36   Calculation of local attention with monotonic alignment.

Figure 2.36 summarizes the operations involved in computing monotonic local attention.

2.10.2.3 Self-Attention Model

This model focuses on different positions within the same input sequence, where the input layer represents *token embeddings*. While the global and local attention models could be adapted to create self-attention, the self-attention model operates on the same input sequence instead of the target output sequence. In this case, the *encoder* determines the score for each *token* within a set of *tokens* associated with the rest of the same set.

Self-attention results in a weighted vector representation based on the similarity (i.e., *dot product*) between pairs of input *tokens* (X) through the attention function. This representation captures relationships between elements of the input sequence, as shown in Figure 2.37.

Typically, an auto-attention mechanism consists of four components:

1. **Query** (Q): It represents the current input (the *"query"*) as a vector, describing the current word or *token* being compared to all other words in the sequence.

2. **Keys** (K): They serve as *labels* for each *token* in an input segment (the "key"). This vector describes what each *token* contributes or when it might be important. The keys should be chosen to identify *tokens* of interest based on the *query*.

3. **Values** (V): These are representations of the actual words (the "values"). Once the relevance of each word is determined, these values are summed to represent the current word.

4. **Attention score**: To determine which *tokens* to focus on, an attention score must be computed. This function takes a *query* and a key as input and produces the attention weight for the *query*-key pair, typically using simple similarity measures such as the dot product. Figure 2.38 presents an example of an attention score matrix computed via a SoftMax function, which assesses the similarity (dot product) between *tokens* in the sequence *"John bought a book."*

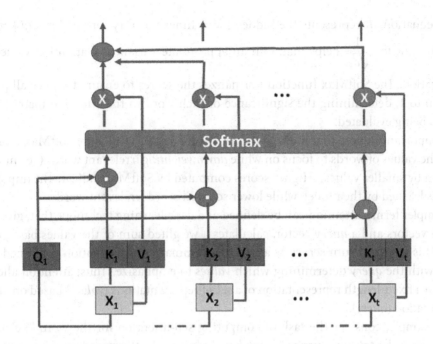

FIGURE 2.37   Self-attention model.

	John	bought	a	book
John	0.8	0.1	0.05	0.05
bought	0.1	0.6	0.2	0.1
a	0.05	0.2	0.65	0.1
book	0.2	0.1	0.1	0.6

FIGURE 2.38   Attention scores for the sentence "John bought a book."

The attention score is computed using pairs (q_i, k_j), where k_j represents a "key" vector of the same dimension as q_i and is linearly connected to the input x_j.

To formalize this, the model components are defined as follows:

$$X = [x_1 x_2 \ldots x_t]^T$$

$$Q = [q_1 q_2 \ldots q_t]^T = X\, W_q \quad K = [k_1 k_2 \ldots k_t]^T = X\, W_k \quad V = [v_1 v_2 \ldots v_t]^T = X\, W_v$$

Here, W_q, W_k, and W_v are the learned weight parameters for Q, K, and V, respectively, while X represents the matrix of input vectors, which are essentially *embeddings* for each *token*. The attention score for these three components (Q, K, and V) is derived by calculating dot products between *keys* and *queries*, followed by a SoftMax function:

$$\text{Atención}(Q, K, V) = \text{SoftMax}\left(\frac{Q\, K^T}{\sqrt{d_k}}\right) * V$$

In this equation, d_k represents the hidden layer dimensionality for *queries and keys*, and the scaling factor $\frac{1}{\sqrt{d_k}}$ helps maintain an appropriate variance of attention values after initialization. The SoftMax function normalizes the scores to ensure they are all positive and sum to 1, determining the significance of each word in the attention matrix for the position being evaluated.

It's important to note that multiplying each value vector (V) by the SoftMax score preserves the values of words to focus on while *downweighting* irrelevant words (i.e., multiplying them by smaller values). Higher scores computed by SoftMax indicate the importance of words learned by the model, while lower scores discard irrelevant words.

In simpler terms, attention can be defined as a deep learning technique that, given a set of *value* vectors and a *query* vector, calculates a weighted sum of the values based on the *query*. This weighted sum serves as a selective summary of information contained in the *values*, with the *query* determining which *values* to emphasize. Thus, attention allows us to obtain a fixed-length representation of a set of representations (*values*) based on another representation (*query*).

For example, consider the task of computing self-attention for the word "he" (a pronoun) within the *token* sequence "*A boy bought books.*" In this scenario, we aim to determine to whom the pronoun "he" refers. To do this, we evaluate each word in the sequence against the *query*. The final score dictates the focus on different parts of the input sentence as we encode a word at a specific position. The score is calculated by taking the dot product of each *query* vector with the corresponding *key* vector of the word we are evaluating. For instance, when processing self-attention for the word at position 1 (i.e., "A"), the first score is the dot product of q_1 and k_1, the second score is the dot product of q_1 and k_2, and so on.

In essence, the *query* can be likened to a sticky note indicating the topic of interest (e.g., "he"), and the *keys* are like labels on folders within a cabinet (representing words in "A boy bought books"). Matching the label to the sticky note allows us to access the contents of that folder, represented by the *value* vector, as illustrated in Figure 2.39. However, it's important to note that we are not merely searching for a single *value*, but rather a combination of *values* from multiple folders. For example, in the given example, "he" has a pronominal link to "a child." Consequently, multiplying the *query* vector by each key vector yields a score for each folder (achieved through the scalar product followed by the SoftMax function), as depicted in Figure 2.40.

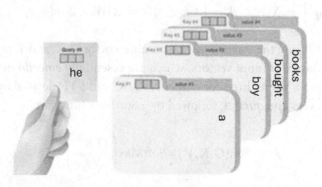

FIGURE 2.39 Self-attention as a search task from a query Q.

These concepts of *query, key, and value* draw an analogy with *information retrieval systems*. For example, when *searching* for a video on YouTube, the search engine maps the *query* to *keys* (video title or description) associated with candidate videos. It then shows the best matching videos (*values*). Self-attention involves scoring each folder (scalar dot product followed by SoftMax) and multiplying each *value* by its score, producing the self-attention layer's output for each word, as illustrated in Table 2.1. The resulting vector can be sent through an FNN to other components of the model.

This weighted combination of value vectors results in a vector that assigns 50% attention to "child," 30% to "a," and 19% to "he."

In general, *K, Q, and V* are matrix representations of *encoder* and *decoder* states, varying depending on the problem. The weight matrices correspond to linear transformations of states and are trained as part of the predictor neural network. Average weights (α) are computed using a SoftMax function over all evaluation function outputs. Higher weights are assigned to vectors whose corresponding keys (*key*) are more similar to the *query* (*query*). This can be described as:

$$\alpha_i = \frac{\exp\left(f_{\text{attention}}\left(\text{key}_i, \text{query}\right)\right)}{\sum_j \exp\left(f_{\text{attention}}\left(\text{key}_j, \text{query}\right)\right)}, \text{out} = \sum_i \alpha_i \cdot \text{value}_i$$

This visualization of attention over a sequence of words is shown in Figure 2.41.

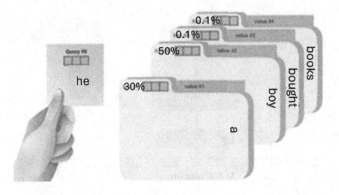

FIGURE 2.40   Self-attention similarity scores with consultation (Q).

TABLE 2.1   Attention Combining to Determine the Referent of "he"

Word	Value Vector	Score	Value Score of X
A		0.3	
boy		0.5	
bought		0.001	
books		0.001	
he		0.19	
		Sum	

2.10.2.4 Multiheaded Attention

To implement a multiheaded attention mechanism, *queries*, *keys*, and *values* are separated into N vectors before applying self-attention. Each "*head*" processes these vectors independently, resulting in N output vectors. These output vectors are then concatenated into a single vector before passing through the final linear layer (see Figure 2.42). This approach allows for different focuses on different parts of the input sequence, improving individual attention performance.

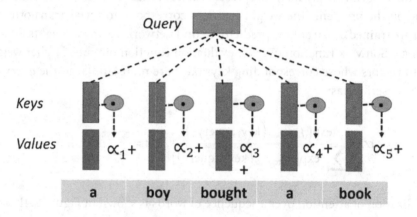

FIGURE 2.41 Visualization of attention over a sequence of words.

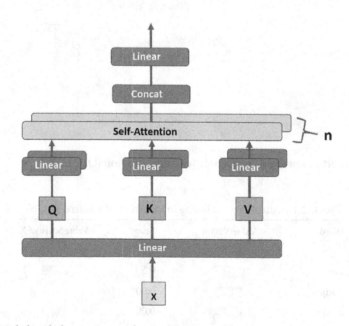

FIGURE 2.42 Multiheaded attention with n attention mechanisms.

Scaled Dot-Product Attention

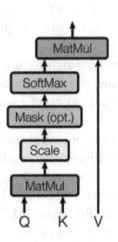

Multi-Head Attention

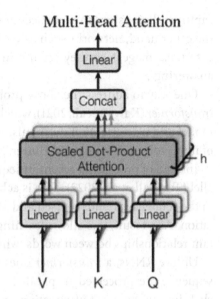

FIGURE 2.43 Optimization of attention calculation blocks for single attention (left) and multi-headed attention (right).

In this case,

$$head_i = \text{Attention}\left(QW^{(i)}_q, KW^{(i)}_k, VW^{(i)}_v\right)$$

So *Attention-Multiheaded* is computed by simply concatenating the heads:

$$\text{Multi-Headed-Attention}(Q, K, V) = \text{Concatenate}(head_1, ...,head_n)W_o$$

where W_o represents the initial weight parameters, and the concatenation is performed by means of the Multiple Sequence Alignment method, which allows aligning three or more sequences of similar length.

A relevant aspect for all the attention models, from the point of view of the performance of all the computational processes previously discussed, is that they are extremely demanding from the point of view of computational resources. This becomes a very critical aspect in LLM, as the matrices being computed are very large. One way to address this problem is to separate the operations into different computational blocks, so that they can be carried out on parallel computing architectures.

For this, one can decompose the matrix multiplication (MatMul), scaling (Scale), SoftMax (SoftMax), and, optionally, *masking* (Mask) computations into different blocks, as shown in Figure 2.43.

2.11 TRANSFORMERS

In general, sequence learning approaches rely on convolution (CNN) or recursion (RNN) to generate word representations, which makes them very inefficient. They are unable to capture relationships in sequences of very long inputs, which limits many applications. On the other hand, networks such as GANs are mainly used for generating realistic synthetic data (i.e., images), so they are not intended for NLP problems (i.e., translation, question answering).

One way to address the above problems is through a deep learning architecture called *transformer* (Kalyan et al., 2021), which is based on the previously described *encoder-decoder* model and uses attention mechanisms to focus on the most relevant aspects of a sequence of input *token*s (Phuong and Hutter, 2022).

Instead of processing *token*s in sequential order, a *transformer* can process them in parallel (Tunstall et al., 2022). This is achieved using attention connections, where each word in the sequence attends to itself and all other words to compute a contextualized representation. As a result, this allows learning contextualized representations that capture important relationships between words, which is key to many NLP tasks (Rothman, 2022).

Unlike RNNs, a *transformer* does not have a cycle structure, so all *token*s in an input sequence are processed in parallel, and the relationship between those *token*s is modeled directly by a self-attenuation mechanism, independent of their respective positions (Cuantum, 2023; Yue, 2023). If we wish to compute the following characterization of a given word, a *transformer* will compare that word one by one with the other words in the sentence and derive the attention score for those words. These scores determine the *semantic* impact of other words in a given vocabulary. The attention score is then used as the average weight for all representations (i.e., *embeddings*) of words, which are provided to a fully connected network, in order to generate a new representation.

Given this implicit parallelism, training a *transformer* is faster than RNN models and performs better on language tasks. Another advantage is its ability to focus on the attentional parts of the network, especially when processing or translating a given word, so that it can understand how information is transmitted through the network.

Let's look at a *transformer* from a high-level view as a black case for an application in an NLP task: machine translation. It takes a sequence of words in one language and produces its translation in another, predicting word-by-word, as shown in Figure 2.44.

At a slightly higher level of detail, a *transformer* is an attention-based *encoder-decoder* type architecture (Ekman, 2022; Tunstall et al., 2022; Vaswani et al., 2017), where the *encoder* transforms an input sequence into a continuous representation (i.e., *embedding*), which maintains all the information learned from such input. Then, the *decoder* takes such representation and, step by step, predicts a simple output (i.e., it always goes predicting by the rightmost word, *shifted right*), according to the most probable *token*, while feeding from the previous output (see Figure 2.45).

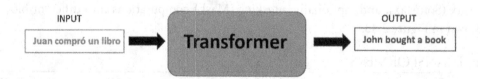

FIGURE 2.44   Overview of a transformer.

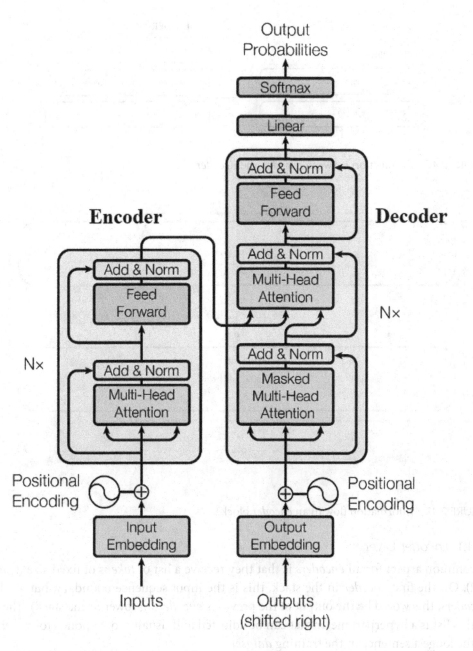

FIGURE 2.45   Typical structure of a transformer architecture.

We can see that the architecture is basically composed of three components: *encoder* (encoding), *decoder* (decoding), and the connections (residuals) between them, which allow the most relevant information to be transferred by the attention mechanisms. The relationship between the different sublayers (*stacks*) of the *encoder* and *decoder* is shown in Figure 2.46.

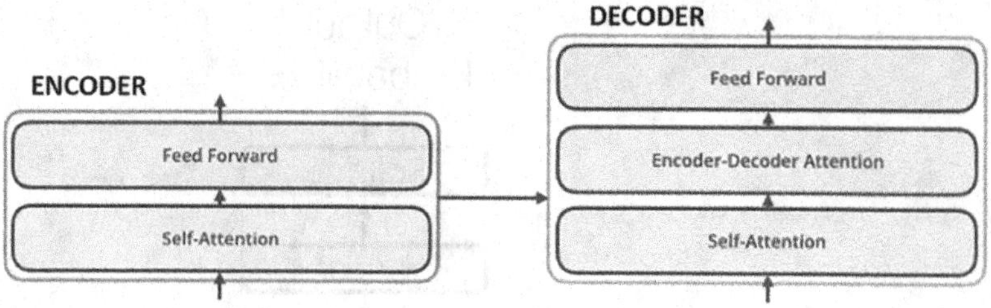

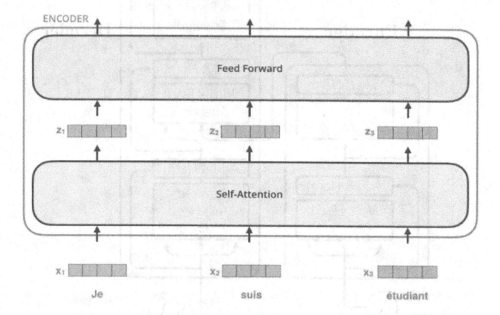

FIGURE 2.46 Connection between *encoder* and *decoder*.

FIGURE 2.47 Information flow in an *encoder* block.

2.11.1 *Encoder* Layer

A common aspect for all *encoders* is that they receive a list of *tokens* of fixed size (usually 512). On the first *encoder* in the stack, this is the input sequence boundary, but, on later *encoders*, this would be the output of the previous *encoder* (or lower in the stack). The size of this list is a hyperparameter that can be adjusted and usually corresponds to the length of the longest sentence in the training *dataset*.

After obtaining the *embeddings* of the words from the input sequence, each of them flows through each of the two layers of the *encoder*, as shown in Figure 2.47.

In the figure, it is shown which word at each input position flows through its own path in the *encoder*. Each path is independent, except for the attention layer, which interconnects them, as described in other sections. Due to their nature, the routes can run in parallel, while flowing through the *feed-forward* layer.

Recall that an *encoder* receives a list of *embeddings* of the input words. Each *encoder* processes this list, passing these vectors to the self-attention layer, then to an FNN, and finally the output is sent to the next *encoder* in the stack.

2.11.2 Positional Encoding

A relevant aspect that has not yet been considered in the model is the way to encode the order of words in the input sequence. In symbolic language models, this *feature* would be encoded directly by the syntactic or semantic structures underlying the sentence. However, in sequence neural models, this is not directly represented. The need for this information is because a *multiheaded* attention mechanism cannot distinguish whether one input comes before another in the sequence or not. In tasks such as language comprehension (Gillon, 2019), position is important for interpreting the input words; for example, an attention-based model might believe that these two sentences possess the same meaning: (1) "John took it away from a dog" and (2) "A dog took it away from John."

This is because there is no notion of word position within a sentence, hence a way to encode not only the *embedding* of words but also the position of words within a sentence is required.

Intuitively, this information or position encoding can be aggregated using *feature* patterns that the neural network can directly capture and potentially generalize to larger sequences.

To achieve this, the *transformer* adds a **Positional Encoding** (PE) vector to each input *embedding*. These vectors follow a specific pattern that the model learns, which allows it to determine the position of each word or the distance between different words in the sequence; that is, the encoding of each word that is input to the first *encoder* is the sum of its *embedding* and its PE.

Up to this point, you are probably thinking, "Why not simply assign a position index to each word in an input sentence?". The main reason for this is that, for long sequences, indexes can grow in magnitude. Normalizing the index value to lie between 0 and 1 can cause problems for sequences of varying length, since they would be normalized differently. On the other hand, a *transformer encoder* does not have recurrence/cycles like RNNs, so we must add information about the positions in the input *embeddings*.

In Figure 2.48, an example of the matrix encoding only positional information is shown. Usually, the computation of the PE vectors is performed as follows:

- For each odd index (i) in the input vector, create a vector using the *cosine* function.

- For each even index in the input vector, create a vector using the *sine* function.

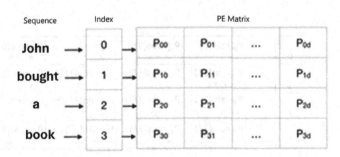

FIGURE 2.48   PE matrix for the sequence "John bought a book."

- The PE vectors are added to the corresponding input *embeddings*. This provides the network with information about the position of each vector.

Formally, PE is represented by the following equation:

$$
PE_{(pos, i)} = \begin{cases} \sin\left(\dfrac{pos}{10\,000^{i/d_{model}}}\right) & \text{if } i \bmod 2 = 0 \\[4mm] \cos\left(\dfrac{pos}{10\,000^{(i-1)/d_{model}}}\right) & \text{else} \end{cases}
$$

where pos is the position of an element in the input sequence, i is an index to represent each of the dimensions and maps the columns of the resulting matrix, and d_{model} is the *embeddings* dimension of the model. In the expression, it can be seen that the even positions correspond to a *sine* function and the odd positions to *cosine* functions. Note that the *sine* and *cosine* functions were chosen together because they have linear properties that the model can easily learn to focus on relative positions. Then, all these concatenated PE values for all dimensions are added to the original input *embeddings*

Intuitively, suppose we draw a *sine* curve and vary pos (on the horizontal axis); we will get different position values on the vertical (y) axis. Therefore, words with different positions will have different values of position *embeddings*. However, since the *sine* curve repeats in intervals, we could have positions with the same values of position *embeddings*, despite being in two very different positions. This is where the parameter i in the equation becomes important. If one varies i in the PE equation, we will get several curves with varying frequencies. Thus, by reading the values of position *embeddings* versus different frequencies, we will get different values in different *embedding* dimensions for different positions.

To understand the expression that computes each PE, let us take an example of the sentence "*John bought a book,*" with $n = 100$ and $d = 4$. In Figure 2.49, the PE matrix for this sentence is shown.

Sequence	index	$i = 0$	$i = 0$	$i = 1$	$i = 1$
John	0	$P_{00}=\sin(0)$ $= 0$	$P_{01}=\cos(0)$ $= 1$	$P_{02}=\sin(0)$ $= 0$	$P_{03}=\cos(0)$ $= 1$
bought	1	$P_{10}=\sin(1/1)$ $= 0.84$	$P_{11}=\cos(1/1)$ $= 0.54$	$P_{12}=\sin(1/10)$ $= 0.10$	$P_{13}=\cos(1/10)$ $= 1.0$
a	2	$P_{20}=\sin(2/1)$ $= 0.91$	$P_{21}=\cos(2/1)$ $= -0.42$	$P_{22}=\sin(2/10)$ $= 0.20$	$P_{23}=\cos(2/10)$ $= 0.98$
book	3	$P_{30}=\sin(3/1)$ $= 0.14$	$P_{31}=\cos(3/1)$ $= -0.99$	$P_{32}=\sin(3/10)$ $= 0.30$	$P_{33}=\cos(3/10)$ $= 0.96$

FIGURE 2.49   PE matrix calculated for the sequence "John bought a book."

To visualize the behavior of this positional information, we could take the positional matrix and plot the *sine* and *cosine* curves with different wavelengths encoding the position in the dimensions, as shown in Figure 2.50.

We can interpret the PE as a sort of "clock," with many hands at different speeds. Toward the end of a PE vector, the hands move slower and slower as the positional index (pos) increases. The hands near the end of the dimensions are slow, because the denominator is large, so the angles are approximately zero there, unless *the* pos is significant enough. Thus, we can imagine that each position has a clock with many hands pointing to a single time.

2.11.3 Residual Connections

Another relevant aspect in the *encoder* architecture is that each sublayer has a *residual* connection around it, and its output goes to a layer normalization (Norm) process. In general, the normalized residual output is projected onto an FNN and corresponds to a pair of linear layers with a ReLU activation function between them. Recall that ReLU (Rectified Linear Unit) is a piecewise linear function that generates the same input if it is positive; otherwise, it generates zero.

Then the output is added back to the input of an FNN and normalized, as shown in Figure 2.51.

These *residual* connections help the training of the network, as they allow the gradients to flow through the network directly. On the other hand, the normalization of the layers is used to stabilize the network, substantially reducing the training time. The visualization of the vectors going through the normalization operation associated with self-attenuation can be seen in Figure 2.52. Note that this also applies to the *decoder* layers.

All these operations make it possible to encode the input into a continuous representation with attention information. In this way, the *decoder* focuses on the relevant *tokens* during encoding. We could stack the *encoder* multiple times (*Nx*), to encode more information, where each layer can learn different attention representations and thus boost the predictive power of the *transformer*.

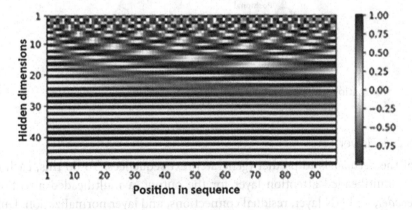

FIGURE 2.50 Positions in a sequence with curves of different wavelengths.

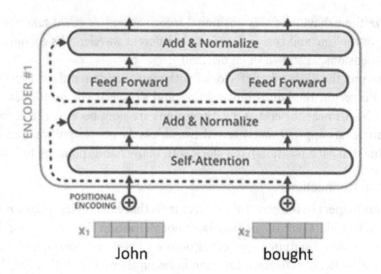

FIGURE 2.51   Residual connections in an *encoder*.

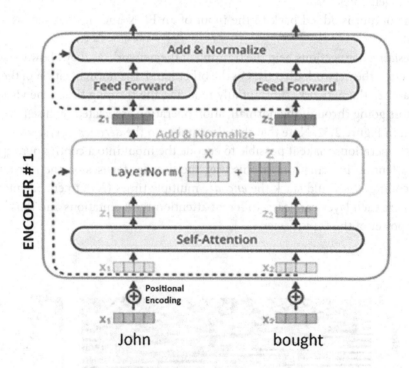

FIGURE 2.52   Summation and residual normalization operation.

2.11.4 *Decoder* Layer

The goal of the *decoder* is to predict (generate) text sequences. To do this, each *decoder* block has a multiheaded attention layer for the output, a multiheaded attention layer, *encoder-decoder*, an FNN layer, residual connections, and layer normalization. Unlike the *encoder*, each *multiheaded* attention layer has a different task. The *decoder* is completed with a linear layer that acts as a classifier, and a SoftMax function to calculate the probabilities of words to be predicted in the output.

The *decoder* is *autoregressive*, so it starts with a start *token* and takes a list of the previous outputs as input, as well as the *encoder* outputs, which contain the input attention information. The *decoder* stops when it generates a *token* as output.

The *encoder* starts processing the input sequence, and the *encoder* output from the top of the stack is then transformed into a set of K and V attention vectors, which are processed by each *decoder* in its *encoder-decoder* attention layer, which allows the *decoder* to focus on relevant positions in the input sequence.

The following steps repeat the process until a special symbol is reached, indicating that the *decoder* has completed its output. The output of each step feeds the lower *decoder* at the next time instant, and the *decoders* accumulate their decoding results just as the *encoders* did. Then, *embedding* is generated, and PE is added to those *decoder* inputs.

The self-attention layers in the *decoder* work in a slightly different way than those in the *encoder*. In the *decoder*, the self-attenuation layer can only focus on positions earlier in the output sequence. This is done by *masking* future positions prior to the application of SoftMax in the self-attention calculation.

The *encoder-decoder* attention layer works like the multiheaded self-attention, except that it creates its array of *queries* (Q) from the layer below it and takes the array of *keys* (K) and *values* (V) from the output of the *encoder* stack.

For example, when calculating attention scores for the word "bought," you should not have access to the word "a," since that word is a future word that was generated later. The word "bought" should only have access to itself and to the previous words, as shown in the attention scores in Figure 2.53. Note that there is a special sequence start *token* called **<start>**.

For this, we need a method that prevents the calculation of attention scores for future words. This method is called *masking*, in which a mask is added before calculating the SoftMax function, and after scaling the values. Thus, the mask is a matrix of the same size as the attention scores filled with 0 and infinite negative (*inf*) values. When a mask is added to the scaled attention scores, a matrix of the scores is obtained, with the upper right diagonal filled with negative infinities, as shown in the example in Figure 2.54.

The reason behind this mask is that, once the SoftMax of masked scores is taken, negative infinities discard zero values, leaving zero attention scores for future *tokens*; for example, in the figure, the scores for am possess values for itself and for all previous words, but it is zero for the word fine. This tells the model that it should not pay attention to such words.

	<start>	John	bought	a
<start>	0.7	0.1	0.1	0.1
John	0.1	0.6	0.2	0.1
bought	0.1	0.3	0.6	0.1
a	0.1	0.3	0.3	0.3

FIGURE 2.53 *Decoder* multi-head attention.

Scaled scores					Forward masks					Masked scores			
0.7	0.1	0.1	0.1		0	inf	inf	inf		0.7	inf	inf	inf
0.1	0.6	0.2	0.1	+	0	0	inf	inf	=	0.1	0.6	inf	inf
0.1	0.3	0.6	0.1		0	0	0	inf		0.1	0.3	0.6	inf
0.1	0.3	0.3	0.3		0	0	0	0		0.1	0.3	0.3	0.3

FIGURE 2.54   Calculation of masked attention scores.

This *masking* is the only difference in how attention scores are computed in the first *multiheaded* attention layer. This layer still has multiple heads to which the mask is being applied, before being concatenated and sent to the linear layer for further processing. The output of the first *multiheaded* attention is a masked output vector with information on how the model should be focused on the *decoder* input.

For the second *multiheaded* attention layer, the outputs of the *encoder* are *queries* and *keys*, and the outputs of the first *multiheaded* attention layer are values. This process matches the *encoder* input to the *decoder* input, allowing the *decoder* to decide which *encoder* input to pay attention to. The output of the second *multiheaded* attention layer feeds the FNN network for further processing.

2.11.5 Linear Layer and SoftMax

Usually, the output of the FNN layer is passed to a final linear layer that acts as a classifier, which is as large as the number of classes you have; for example, if you have 10,000 classes for 10,000 *tokens*, the output of such a classifier will be of size 10,000 or more known as a vector *logits*.

The output of the classifier is fed into the SoftMax layer, which will produce probability values between 0 and 1. Then, the index of the *token* with the highest probability is taken, which corresponds to the words that are predicted at an instant. The *decoder* takes the output, adds it to the *decoder* input list and continues decoding until a *token* is predicted. The highest probability prediction is the final class that is assigned to the final *token*.

Note that the *decoder* can also be stacked in multiple layers (*Nx*), each of which takes inputs from the *encoder* and previous layers. By stacking the layers, the model can learn to extract and focus on different combinations of attention from the attention heads, potentially improving its predictive power.

2.11.6 Training

During the training process, the data consists of two parts:

1. The input sequence (e.g., "You are welcome," for a translation problem).

2. The target sequence (e.g., "De nada" in Spanish).

The structure connecting all these steps is shown in Figure 2.55, where there are some relevant aspects of the architecture:

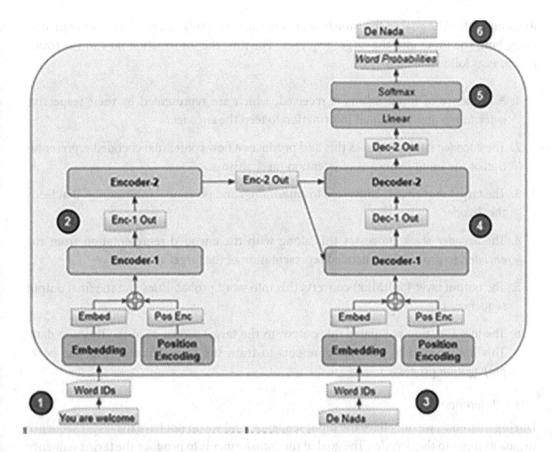

FIGURE 2.55 Training of a transformer.

- One can stack as many *encoder* and *decoder* blocks (*Nx*), as one sees fit. However, this substantially increases the training cost.

- More *encoder* or *decoder* blocks can be added, allowing more nonlinearity to be added to the mapping between input and output, achieving a more powerful model. This is specifically because each block adds layers of FNNs that increase complexity.

- All *encoders* are identical in structure, but do not share neural network weights. Each is decomposed into two sublayers:

  1. *Attention layer:* this is the first stage of the flow and helps the *encoder* focus on other words in the input sentence while encoding a specific word. The output is sent to the next layer.

  2. *Feed-forward neural network (FNN):* It computes the final context vectors and then sends them to the *decoder*.

- The *decoder* has both layers, but between them there is an attention layer, which helps the *decoder* to focus on relevant parts of the input sequence.

Remember that the goal of the *transformer* is to learn how to generate the target sequence, using both the input and the target sequence. For this, the information flow in the *transformer* is as follows:

1. A sequence of input *tokens* is received, which are represented by their respective *embeddings* and positional information to feed the *encoder*.

2. The *encoder* stack processes this and produces a new contextual encoded representation of the input by means of attention mechanisms.

3. The target sequence is converted to *embeddings* and positional information that feeds the *decoder*.

4. The *decoder* stack processes this along with the encoded representation from the *encoder*, to produce an encoded representation of the target sequence.

5. The output layer (SoftMax) converts this into word probabilities and the final output sequence.

6. The loss function compares this output to the target sequence of the training data. This loss is used to generate gradients to train the *transformer* during the *backpropagation* process.

2.11.7 Inference

During inference, we only have the input sequence, and we do not have the target sequence to pass as input to the *decoder*. The goal of the *transformer* is to produce the target sequence only from the input sequence.

Then, the output is generated in one cycle, and we feed the output sequence from the previous time step to the *decoder* in the next time step, until we find an end-of-sentence *token*. The difference with a *Seq2Seq* model is that, at each time step, we re-feed the entire output sequence generated so far, rather than just the last word.

The data flow during inference is as follows (see Figure 2.56):

1. The input sequence is converted into *embeddings* (with position encoding) and sent to the *encoder*.

2. The *encoder* stack processes this and produces an encoded representation of the input sequence.

3. Instead of the target sequence, we use an empty sequence with only a sentence start *token*. This is converted into *embeddings* (with position encoding) and sent to the *decoder*.

4. The *decoder* stack processes this along with the encoded representation from the *encoder* stack, to produce an encoded representation of the target sequence.

5. The output layer converts this into word probabilities and produces an output sequence.

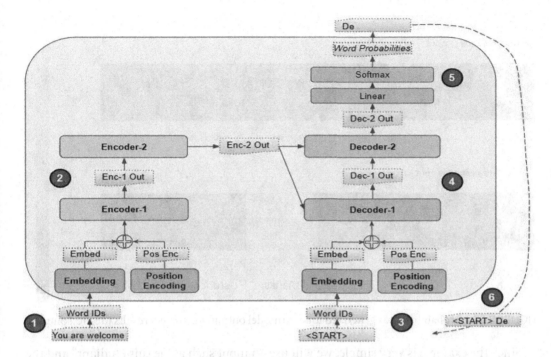

FIGURE 2.56   Inference in a transformer.

6. We take the last word of the output sequence as the predicted word. That word now fills in the second position of our *decoder* input sequence, which now contains a sentence start *token* and the first word.

7. Return to step (3). Provide the new *decoder* sequence in the model. Then, take the second word from the output and add it to the *decoder* sequence. Repeat this until you predict an end-of-sentence *token*. Note that since the *encoder* sequence does not change for each iteration, we do not have to repeat steps (1) and (2) each time.

2.11.8 Loss Function

The *loss* function during the training stage is the metric that the model is optimizing to arrive at a more accurate model.

Suppose we are training the model with a simple example: translating *merci* into "thank you." This means that we want the output to be a probability distribution indicating the word "*thank you*." However, since this model is not yet trained, that is unlikely to happen yet, so a difference occurs, as shown in Figure 2.57, when comparing the probability distributions for each word in a vocabulary of six *tokens*. These distributions are usually compared by subtracting one from the other, via cross-entropy or KL divergence.

Recall that this is because, in a neural network, the model parameters (weights) are randomly initialized, so the untrained model produces a probability distribution with arbitrary values for each *token*. The network weights are error-corrected using *back-propagation* algorithms to produce at each instant an output closer to the desired result.

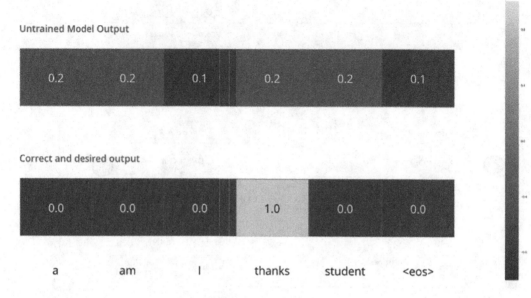

FIGURE 2.57   Probabilities for generated tokens (model output) vs. correct results (desired output).

Since the example is very simple, we will use an input such as "je suis étudiant" and the expected output: "I am a student." What this really means is that we want our model to successively generate probability distributions where:

- Each probability distribution is represented by a vector of width equivalent to the size of the vocabulary (i.e., 6 in our example but, more realistically, a number like 30,000 or 50,000).

- The first probability distribution has the highest probability in the cell associated with the word "I."

- The second probability distribution has the highest probability in the cell associated with the word "I am."

- And so on, until the fifth output distribution indicates the symbol **<eos>**, which also has an associated cell of the 10,000-item vocabulary.

After training the model on a sufficiently large *dataset*, we expect the probability distributions produced to be better able to predict the correct words in the output sequence.

Since the model produces the outputs (i.e., words) one at a time, there are two popular search strategies to select the best word (Freitag and Al-Onaizan, 2017):

- *Greedy Decoding*: the model selects the word with the highest probability from that probability distribution and discards the rest.

- *Beam Search*: the model keeps in memory the top *n* words (e.g., "I" and "a") and, in the following steps, the versions of the words that produce less prediction error are kept.

2.12 CONCLUSIONS

In this chapter, we explored the basics of language models, including neural networks, sequence models, attention mechanisms, *encoder-decoder* architectures, and *transformers*.

Sequence models (*Seq2Seq*), such as RNNs, short-medium term networks (LSTMs), and CNNs, have been widely used for language modeling. These models capture temporal and structural dependencies in textual data, allowing them to generate well-formed output sequences.

One capability that will not be considered by traditional neural models is attention mechanisms, which play a crucial role in enhancing the ability of language models to capture long-term relationships between words. Attention enables models to assign different importance weights to input words, allowing them to focus on the most relevant parts of the context.

Transformers represent a significant innovation in the field of language models for processing large sequences of input. These architectures use attention mechanisms and *encoder-decoder* architectures, enabling the generation of coherent, high-quality text sequences.

## CHAPTER 3

Large Language Models

3.1 INTRODUCTION

Language modeling has been extensively studied for language understanding and generation in the last two decades, evolving from statistical to neural models. Recently, *pre-trained* language models using *transformers* on large-scale corpora have been proposed, showing strong capabilities for solving various NLP tasks (Phuong and Hutter, 2022). This scale can lead to improved performance, so the scaling effect has been further studied by increasing the model size to an even larger size. When the parameter scale (i.e., number of weights to fit in a neural network) exceeds a certain level, these scaled-up LMs not only achieve significant performance improvement, but also show some special abilities that are not present in small-scale LMs (Ge et al., 2023). To discriminate the difference in parameter scale, the expression large language models (Large Language Model) or LLM has been used for PLMs of significant size (Zhao et al., 2023). Significant progress has been made in LLM research in recent times, and a notable progress is the release of applications such as ChatGPT, which has attracted wide attention from society (Wake et al., 2023).

LLMs usually refer to LMs containing hundreds of billions of parameters, which are trained on massive text or corpus data, and heavily based on *transformer* architectures, containing many blocks of *encoders*, and producing very deep neural networks (Mialon et al., 2023). LLMs, to a large extent, scale the model size, training data, and total computation, so they can better understand natural language and generate high-quality text based on the given context. However, some abilities (i.e., learning in context) are unpredictable and can be observed only when the model size exceeds a certain level (Bommasani et al., 2021).

In this chapter, common *feature s* of LLMs such as emergent skills, data used, types of training, types of learning, and types of *tokenization* are introduced. Based on this, the main LLMs and their differentiating aspects are then described.

3.1.1 Emergent Skills

The emergent skills of LLMs correspond to certain capabilities that are not present in small models, but emerge in LLMs and that distinguish them from previous pre-trained models. Moreover, their performance increases significantly when the scale reaches a certain level

DOI: 10.1201/9781003517245-3

(Boiko, MacKnight, and Gomes, 2023). Currently, there are at least three representative types of emerging skills:

1. **Context learning**: the context-learning ability is formally introduced in GPT-3. Assuming that the model has received natural language instruction and/or several task demonstrations, it can generate the expected output for test instances by completing the sequence of words in the input text, without the need for additional training or gradient updates.

2. **Instruction tuning**: by *fine-tuning* a model and a combination of multi-task corpora formatted through natural language descriptions (instructions), LLMs perform well on tasks not seen in the form of instructions. With this capability, instruction *fine-tuning* (i.e., *instruction tuning*) enables new tasks by understanding task instructions without using explicit examples (Peng et al., 2023).

3. **Multi-step reasoning**: for small models, it is difficult to solve complex tasks involving multiple reasoning steps such as mathematical-verbal problems, whereas, with a "chain-of-thought" reasoning strategy, LLMs can solve such tasks using *prompts* or descriptions of the input, usually known as *prompts*. This enables intermediate reasoning steps that allow the final answer to be derived (Prystawski and Goodman, 2023).

Depending on the size of the models, some emergent skills may include[1]:

- **Small (8B–13B parameters)**: basic arithmetic, code debugging, reading comprehension, and basic language operations (e.g., creating stories, essays, poetry).

- **Medium (64B–175B parameters)**: language puzzles, understanding and solving college admissions tests, physical intuition, logical deduction, or understanding metaphors.

- **Large (70B–540B parameters)**: geometric shapes, phonetic alphabet, causality, elementary mathematics, and code explanation.

- **Extra-large (1 teraparameter)**: spatial reasoning, creativity, app construction, picture handling, and thematic test performance.

3.1.2 Skills Enhancement Techniques

To develop LLM, usually a number of important techniques that significantly improve their capability, are used, which include the following (Puchert et al., 2023):

- **Scaling**: this is the key factor in increasing the capacity of LLMs. A large model size is essential for emerging skills. However, scaling is performed not only on the model size, but also on the *dataset* size. For this, there are usually three key aspects: model size, data size and total computation.

- **Training**: due to the huge size of the models, it is very difficult to successfully train an LLM, as efficient training algorithms are needed to learn the network parameters, several times executed in parallel. Due to this, several optimization approaches have been proposed to facilitate the implementation and deployment of parallel algorithms. Recently, the development of GPT-4 used special infrastructure and optimization methods that reliably predict LLM performance with much smaller models.

- **Elicitation**: after pre-training an LLM on large-scale corpora, they can potentially solve general tasks. This type of ability may not be explicitly exhibited when LLMs perform some specific tasks. However, it is useful to design appropriate task instructions or context-specific strategies to elicit such skills; for example, a chain-of-thoughts strategy may be useful for solving complex reasoning tasks by including intermediate reasoning steps.

- **Alignment adjustment**: since LLMs can capture data *feature s* from pre-training corpora, it is likely to generate toxic, biased or even harmful content for humans. Because of this, it is necessary to align LLMs with human values; e.g., useful, honest, and harmless. To this end, models trained with *instruction tuning* such as InstructGPT (Wang et al., 2022; Peng et al., 2023) allow an LLM to follow expected instructions using *reinforcement learning* with human feedback (RLHF) techniques, in the training cycle (Christiano et al., 2023).

3.1.3 Corpora

LLMs consist of a significantly larger number of parameters that require a larger volume of training data covering a wide range of content. For this purpose, there are increasingly accessible *datasets* or training corpora that have been released for research such as the following:

- **Books**: these include BookCorpus, which consists of over eleven thousand books of various subjects and genres, and *Project Gutenberg*, which consists of over seventy thousand literary books (i.e., novels, essays, poetry, drama, history, science, philosophy...).

- **CommonCrawl**: it is a *web crawling dataset* containing petabytes of data that has been widely used as training data for LLMs.

- **Reddit**: it is a social networking platform that allows users to submit links and text posts, which others can vote on through "upvotes" or "downvotes." Highly voted posts are often considered useful and can be used to create high-quality *datasets*.

- **Wikipedia**: it is an online encyclopedia that contains a large volume of high-quality articles on a variety of topics. Most of these articles are composed in an expository writing style, covering a wide range of languages and areas.

- **Code**: to collect code data, open source code is usually extracted from the Internet through GitHub and Stack Overflow. In addition, Google has released public *datasets* such as Big *Query*, which contains many open source licensed code snippets in various programming languages.

- **Others**: this includes the open source Pile *dataset*, which contains over 800 GB of data from multiple sources, including books, websites, code, scientific articles, and social networking platforms. The *dataset* is constructed from 22 high-quality, diverse subsets.

3.1.4 Types of Training

An important aspect of LLMs is the type of technique used for their training which, in general, includes the following:

- **Pre-training**: it refers to the task of training an LM on a large and diverse *dataset* before adjusting it to a specific task. During pre-training, the model is trained to learn general language *features*, such as semantics and syntax, which can then be used for specific tasks; for example, a model can be pre-trained on a general text set (i.e., Wikipedia) and then fit the model to a specific task, such as news text generation.

- **Fine-tuning**: it refers to the task of continuing to train a pre-trained language model on a smaller, task-specific *dataset*. *Fine-tuning* or *fine-tuning* is useful when a limited *dataset* is available and one wishes to improve the performance of the model on a specific task; for example, one can take a pre-trained model on general text and tune it to a movie subtitle generation task (Howard and Ruder, 2018).

- **Instruction tuning**: this is an approach for *fine-tuning* pre-trained LLMs over a set of instances formatted in natural language form. To perform *fine-tuning*, instruction-formatted instances (i.e., input, indication, or output) must be collected or constructed. Then, these formatted instances must be used to tune LLMs in a supervised learning fashion.

3.1.5 Types of Learning

In general, there are four common machine learning techniques for LLMs:

- **Semi-supervised learning**: this training paradigm combines unsupervised pre-training with supervised *fine-tuning*. The goal is to train a model with a large unsupervised *dataset*, then fine-tune the model to different tasks by using supervised training on smaller *datasets*.

- **Zero/one/some-shot learning**: in general, deep learning systems are trained and tested for a specific set of classes. If a document categorization system is trained to classify descriptions of cats, dogs, and horses, it could only be tested on those three classes. Conversely, in zero *shot*) learning environments, in testing the system, it is shown, without updating the weights, classes that it has not seen at the time of training (i.e., testing the system on elephant descriptions). The same is true for the *one-shot* and *multi-shot* settings, but, in these cases, at the time of testing, the system sees one or a few examples of the new classes, respectively (Brown et al., 2020).

- **Multi-task learning**: most deep learning systems are single-task (e.g., AlphaZero). Multi-task systems overcome this limitation, as they are trained to solve different tasks given a given input; for example, if the word "gato" is entered into the system, it could be asked to look up the Spanish translation of "gato," display the image of a cat, and/or describe its characteristics.

- **Zero/one/few-shot task transfer**: this approach combines the concepts of zero/one/ few-shot learning with multi-task learning. Instead of showing the system new classes at test time, we could ask it to perform new tasks (i.e., showing it zero, one, or a few examples of the new task); for example, assume a system trained on a large corpus. In a one-try *task transfer environment*, we might write, "I love you → Te quiero. I hate you → ___." Implicitly, we are asking the system to translate a sentence from English to Spanish (a task in which it has not been trained) by showing it a single example (one shot).

3.1.6 Types of Tokenization

In the context of NLP tasks, *tokenization* refers to the way in which a text segment is represented as a sequence of vocabulary items, usually called *tokens* (Kudo and Richardson, 2018). Although, in general, various LMs perform *tokenization* at the word level, this is not always the case and depends on the level of granularity desired in a task. Suppose we wish to *token*ize the sentence "We want a complete pre-report"; the usual types of *tokenization* would be as follows:

- **Character-level *tokenization***: the vocabulary V is the alphabet of the language (i.e., English) plus punctuations. In the example sentence, we would obtain a sequence of length 31: ['Q', 'u', ' ', ...]. Usually, this type of *tokenization* tends to produce very long sequences.

- **Word-level *tokenization***: the vocabulary V is set of all the words of the language in question, plus punctuations. In the example sentence, we would get a sequence of length 4: ['We want', 'a', 'pre-report', ...]. Usually, this type of *tokenization* requires a very large vocabulary and cannot handle new words at test time.

- **Subword *tokenization***: the vocabulary V is a set of common word segments such as 'pre,' 'mos,' 'pre.' Common words such as 'a' are usually *token*ized, and individual characters are also included in V, to ensure that all words can be expressed.

Once the vocabulary items are *token*ized, a unique index is assigned. Then, special *tokens* are added to the vocabulary. The number of special *tokens* varies, but usually three are considered: (1) mask *token* (MASK), used in masked language modeling; (2) beginning-of-sequence *token* (BOS), and (3) end-of-sequence *token* (EOS). The complete vocabulary has $|V|$ elements. Thus, a text fragment is represented as a sequence of indices (i.e., *token* IDs) corresponding to its (sub)words, preceded by BOS and followed by EOS.

3.2 BERT

NLP models based on deep learning techniques require very large amounts of data to improve their performance when trained on millions or billions of annotated training examples. To help bridge this gap, several techniques have been developed to train general-purpose language representation models, better known as "pre-training," which uses large corpora of unannotated text. These pre-trained models can be tuned to smaller task-specific *datasets*; e.g., question-answer systems, sentiment analysis, text prediction, text generation, etc.

This approach shows significant improvements in accuracy compared to training on smaller task-specific *datasets* from scratch. One of the models addressing these difficulties in pre-training for NLP tasks is called BERT (Bidirectional *Encoder* Representations from *Transformers*) (Devlin et al., 2019).

BERT is a model that is trained bidirectionally for prediction tasks. This means that we can now have a deeper sense of context and language flow compared to unidirectional LMs (Devlin et al., 2019). BERT is based on a basic *transformer*, consisting only of *encoders*, to read text input, and a *decoder*, to produce a prediction for a task. This *transformer* applies attention mechanisms to understand the relationships between all words in a sentence, regardless of their respective positions.

Generally, BERT can be used to extract high-quality language *features* from a corpus or to fit these models to a specific task (e.g., question answering) with proprietary data in order to produce good predictions.

As an LM predictor, BERT's goal is to "fill in the blank," depending on the context; e.g., given.

"John bought a _____ at the corner store."

An LM could complete this sentence by saying that the word "vehicle" would fill in the blank 20% of the time and the word "eggs" 80% of the time.

Instead of predicting the next word in a sequence, BERT uses a technique called "masked language model" or MLM, which randomly *masks* words in a sentence and then attempts to predict them. *Masking* means that the model looks in both directions and uses the entire context of the sentence, both left and right (aka bidirectional), to predict the masked word (i.e., that "hidden" or missing word). Thus, BERT takes into account both the previous and the next *token* at the same time, hence *bidirectional*.

Unlike context-independent models such as Word2Vec, the above *feature* allows BERT to generate context-based representations for each word based on the other words in the sentence; for example, in the sentence "I accessed the bank account," a one-way contextual model would represent "bank" based on "I accessed" but not "account." In the case of BERT, the word "bank" is represented using both its preceding and following context-"I accessed the... account"-starting from the bottom of a deep neural network, making it bidirectional.

3.2.1 Operation

Since the goal of BERT is to generate a language representation model, it only uses the *encoder* component of the *transformer* (Figure 3.1).

In training, the model receives pairs of input sentences, and it must learn to predict whether the second sentence of the pair is the next sentence (*IsNext*) in the original text. During training, 50% of the inputs are a pair in which the second sentence is the next sentence in the original text, while, in the other 50%, a random sentence is chosen from the corpus. The assumption is that the random sentence will be disconnected from the first sentence. To achieve this, the training uses two strategies:

Masked LM (MLM): 15% of the words in the input are randomly masked (hidden), replaced with a *token* [**MASK**], and the entire sequence is processed through the *encoder*. The model then attempts to predict only the masked words, based on the context provided by the other unmasked words in the sequence. The model only attempts to predict when the [MASK] *token* is present in the input, whereas it is desired that the model predict the correct *token*s, regardless of which *token* is present. To address this, of the 15% of the *token*s selected for *masking*:

- 80% of the *token*s are replaced with the masked *token* [MASK].

- 10% of the time, *token*s are replaced with a random *token*.

- 10% of the time, *token*s are left unchanged.

Then, the prediction of the output words requires:

- Adding a classification layer on top of the *encoder* output.

- Multiplying the output vectors by the *embeddings* matrix, transforming them into the vocabulary dimension.

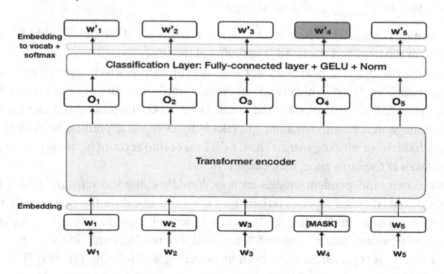

FIGURE 3.1 *Encoder*-based architecture of BERT.

- Calculating the probability of each word in the vocabulary with the SoftMax function.

  - **Next Sentence Prediction (NSP)**: to understand the relationship between two sentences, the training process also uses NSP. During training, the model obtains, as input, pairs of sentences and learns to predict whether the second sentence is also the next sentence in the original text. Thus, the training of the model is fed with two input sentences at a time, so that:

    - 50% of the time, the second sentence comes after the first sentence.

    - 50% of the time, the second sentence is a random sentence from the entire corpus.

Thus, NSP allows the model to learn about the relationships between sentences by predicting whether a given sentence follows the previous sentence or not; for example, given two sentences A and B, is B the next real sentence that comes after A (isNext) or simply a random sentence in the corpus (NotNext)? The model should predict as follows:

```
Sentence A: the man went to the store
Sentence B: he brought a bottle of milk
Label (prediction): IsNext

Sentence A: the man went to the store
Sentence B: mice don't fly
Label (prediction): NotNext
```

To help the model distinguish between the two sentences in training, the input should be preprocessed as follows, before feeding the model:

- *Token embeddings*: a classification *token* [CLS] is added to the input *token*s (words) at the beginning of the first sentence, and a *token* [SEP] (separation) is inserted at the end of each sentence.

- *Embeddings* **of segments**: a marker indicating whether it corresponds to sentence A or sentence B is added to each *token*. This allows the *encoder* to distinguish between sentences.

- **Positional** *embeddings*: a positional *embedding* is added to each *token* to indicate its position in the sequence.

Thus, to learn to predict whether sentence B is actually connected to sentence A, the following steps are performed:

- Apply the entire input sequence to the *transformer*.

- Transform the output of the token [CLS] into a vector, using a simple classification layer.

- Calculate the probability of IsNext with SoftMax.

- Simultaneously train MLM and NSP in order to minimize the combined loss function of the two strategies.

Then, this model is trained according to the desired number of processing layers on a large corpus (i.e., Wikipedia+BookCorpus) for a long time (one million update steps). Once trained, BERT can be used as a pre-trained model or by *fine-tuning* to specific domains.

3.2.2 Architecture

BERT uses a *transformer* that successively processes an input sequence through a stack of *encoder* layers. In general, there are two types of pre-trained versions of BERT, depending on the scale of the model architecture:

- *BERT-Base*: 12 layers (blocks) of *encoders*, 768 nodes hidden in the FNN, 12 attention heads, 110M parameters.

- *BERT-Large*: 24 layers of *encoders*, 1024 nodes hidden in the FFN, 16 attention heads, 340M parameters.

3.2.3 Model Input

The first input *token* is provided with a [CLS] *token*. Like the basic *transformer encoder*, BERT takes a sequence of *tokens* as input, which flows up the *encoder* stack. Each layer of the *encoder* applies the usual self-attention and passes its results through an FNN network that passes them to the next *encoder* in the stack.

3.2.4 Model Output

Each input position generates an *embedding* (i.e., size 768 in BERT-Base) that can be used to perform *fine-tuning* of the model for different specific tasks such as question answering or sentence classification. For this, only a single layer must be added to the main model:

1. *In text classification tasks*, they are performed similar to NSP by adding a classification layer on top of the *transformer* output for the [CLS] *token*; for example, for a task such as binary sentence classification, the output is just the first position containing the [CLS] *token* (Figure 3.2).

2. *In question-answering or QA tasks*, the model receives a question about a sequence of text and must mark the answer in the sequence. With BERT, a question-answer model can be trained by learning two additional vectors that mark the beginning and the end of the answer.

3. *In Named-Entity recognition (NER) tasks*, the model receives a sequence of text and must mark the different types of entities that appear in the text. With BERT, a NER model can be trained by feeding the output vector of each *token* into a classification layer that predicts the NER label.

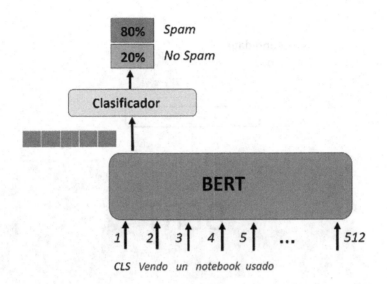

FIGURE 3.2   BERT setting for binary classification tasks.

4. *In MLM tasks*, beyond *masking* 15% of the input, BERT sometimes randomly replaces a word with another word and asks the model to predict the correct word in that position (Figure 3.3).

5. *For sentence prediction tasks*, for example, given a Wikipedia entry as an entry, and a question about that entry as another entry, can we answer that question? To accomplish this, BERT pre-training includes the task of, given two sentences (A and B), is B likely to be the sentence following A, or not (Figure 3.4)?

6. *For a feature extraction task (embeddings)*, the pre-trained BERT model can be used to obtain the contextual *embeddings* of each word. These can then be used as inputs to other models that require *feature* representation (i.e., classifiers, NER).

3.2.5 BERT-Based Pre-Trained Models

There are several BERT-based pre-trained models available including:

- **RoBERTa**: it uses a training approach based on dynamic *masking* and a more intensive training strategy than BERT to learn, more effectively, from data. It was trained on a larger *dataset* than BERT and used a *data augmentation* technique called "random data retraining" to improve the robustness of the model (Liu et al., 2019).

- **ALBERT**: this model is a lighter and more efficient version of BERT, which uses model compression techniques to reduce the size and computational cost.

- **DistilBERT**: this is a smaller and faster model of BERT that uses distillation techniques to transfer knowledge from BERT to a smaller and more efficient model.

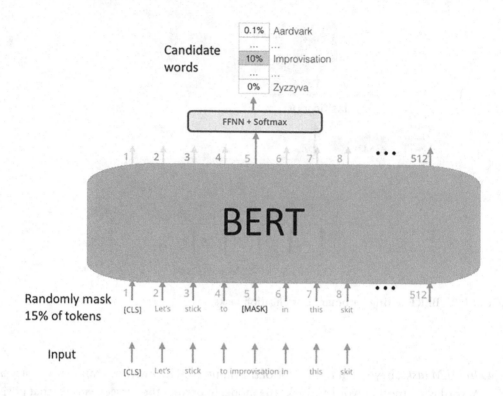

FIGURE 3.3   BERT tuned for mask prediction tasks.

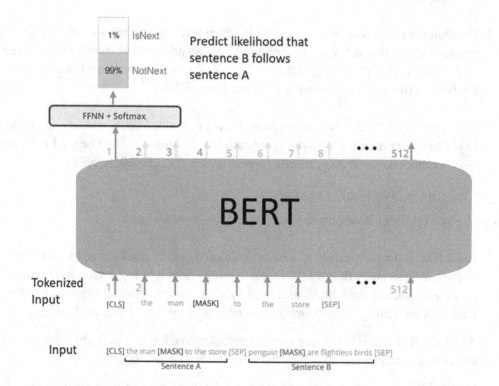

FIGURE 3.4   BERT tuned for next sentence prediction (NSP).

- **ELECTRA**: this model uses a discriminative training approach to improve the efficiency of BERT, which makes it more efficient than NSP, allowing it to be trained on larger and more complex *datasets*.

- **SciBERT**: is a model designed specifically for science and technology related tasks. It is pre-trained on a corpus of scientific text, including journal articles, patents, and other technical documents.

- **FinBERT**: is a model designed for finance-related tasks and is pre-trained on a financial *dataset*, including news and press releases, allowing it to understand and analyze financial language more accurately.

3.3 GPT

A limitation of the BERT model for various tasks is that it is not a generative LLM per se. This means that it cannot generate new text. On the other hand, although BERT is a bidirectional model, it only uses the preceding and following information in one direction during training, which limits its ability to capture the long-distance relationship between words. In addition, BERT requires a fixed-length input (i.e., 512), which means that it cannot process variable-length inputs, which can become a limitation for some applications. Finally, to use BERT for a specific task, *fine-tuning* of the model in that task is required, which can be costly in terms of time and resources.

Intuitively, one way to address these difficulties is through generative LMs that must possess two basic characteristics:

1. *To pre-train an LLM using very large corpora.*

2. *To adapt the pre-trained model to solve specific tasks.*

To address the above, the generative model called GPT (Generative Pre-trained *Transformer*) emerges, which is a large-scale unsupervised LLM trained to predict the next word (Dai et al., 2022). Specifically, GPT stands for the following (Kublik and Saboo, 2022):

- **Generative**: the model was trained to predict (or generate) the next *token* in a sequence of *tokens*, unsupervised. Thus, the underlying relationships between variables in a *dataset* can be learned to generate new data samples similar to those in the original *dataset*. In other words, the model is provided with a large amount of raw text data and is asked to find text *feature s* to create more text.

- **Pre-trained**: it is a language model that has been trained on large *datasets*. This allows them to be used for tasks where it would be difficult to train a model from scratch. A pre-trained model can avoid reinventing the wheel, save time and improve performance. The model is then tuned for specific tasks, such as machine translation.

- **Transformer**: it uses the *transformer* architecture that uses deep *encoder-decoder* models and attention mechanisms. This is a model designed to handle sequential data, such as text.

Different versions of the model have been generated. In version 2 (GPT-2), the model was trained on 40 GB of texts available on the Internet and adjusted to 1500M parameters (Radford et al., 2018). To ensure the quality of the texts used, only Internet pages that have been selected by and rated by humans (i.e., Reddit) were used.

In general, the way this type of model operates is based on a so-called *autoregressive* approach: after each *token* is generated, it is added to the input sequence, and this new sequence becomes the input of the model in its next step.

A key *feature* of this type of generative LLM is that it possesses the ability to generate high-quality coherent text samples. On the other hand, GPT-2 outperforms other LLMs trained on specific domains without the need to use domain-specific *datasets*. In tasks such as question-answering, reading comprehension, summary generation, and machine translation, GPT-2 can learn without using task-specific training data.

3.3.1 The GPT and GPT-2 Models

The basic intuition behind GPT and GPT-2 is the use of generic, pre-trained LLMs to solve a variety of language modeling tasks with high accuracy. In simple terms, this means training the model by *(i)* sampling some text from a *dataset* and *(ii)* predicting the next word. This procedure is a form of self-supervised learning, as the next correct word can be predicted simply by looking at the next word in the *dataset*.

To achieve the above, the GPT model introduces three innovations:

1. **Task conditioning**: the LLM training objective is formulated as *P(output|input)*. However, GPT-2 learns multiple tasks using the same unsupervised model. To achieve that, the learning objective is reformulated as *P(output|input, task)*. This modification, known as "task conditioning," expects the model to generate different outputs for the same input for different tasks. Some models implement task conditioning at an architectural level, where the model is fed by both the input and the task. Therefore, task conditioning is performed by providing examples or natural language instructions to the model to perform a task, which forms the basis for *zero-shot* task transfer (i.e., zero shot or examples).

2. ***Zero-shot* learning and task transfer**: a capability of GPT 2 is *zero-shot* task transfer. Instead of rearranging the sequences, as in previous versions, the input to GPT-2 is provided in a format that allows the model to understand the nature of the task and provide responses. This allows emulation of *zero-shot* task transfer behavior; for example, for the English to French translation task, the model is given an English sentence followed by the French word and a message (:). Then, the model is supposed to understand that this is a translation task and, therefore, to give the French counterpart of the English sentence.

3. **Architecture**: it has 1500M parameters, which is 10 times more than GPT (117M). In addition, GPT-2 has 48 *layers* and uses *embeddings* of 1600 dimensions, a larger vocabulary of 50,257 *tokens*, a larger context window of 1024 *tokens* and a normalization layer after the final self-attention block.

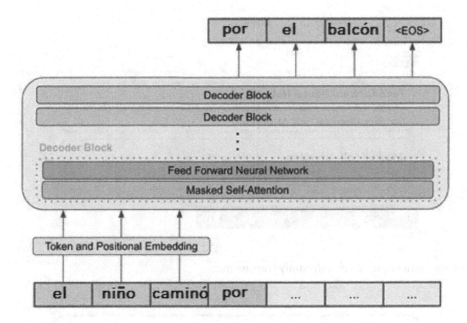

FIGURE 3.5   *Decoder*-only transformer architecture.

Both GPT and GPT-2 use a *decoder*-only *transformer* architecture, allowing the following *transformer* components to be eliminated: each *decoder* layer simply consists of a masked self-attention layer followed by an FNN model. By stacking several of these layers, a deep *decoder*-only *transformer* architecture (*decoder*-only *transformer*) is formed, as shown in Figure 3.5.

Intuitively, think about what the *decoder* is actually doing: given an encoded representation of an input sequence, it must generate a new sequence of words (*w*) in an instant:

$$W_t = Decoder(W_{t-1})$$

This is precisely what an LLM is supposed to do. The reason for using only the *decoder* layers is because the self-attention layers masked within the *decoder* ensure that the model cannot look ahead in a sequence when creating a *token* representation. In contrast, the bidirectional self-attention used in *encoders* allows the representation of each *token* to adapt based on all other *tokens* within a sequence.

Unlike BERT, the self-attention layer masks future *tokens*, interfering with the self-attention computation and blocking the information of *tokens* that are to the right of the position being computed; for example, if we are going to highlight the path of position #4, we can see that only present and previous *tokens* are allowed to be paid attention (aka attended) (Figure 3.6).

A normal self-attention block allows, in one position, to look at the *tokens* to the right. In this case, it is required to prevent that, so masked self-attention is required, as the model should not look ahead in the sentence while predicting the next *token*. Thus, the self-attention mechanism allows the model to capture various *attention patterns*: sometimes,

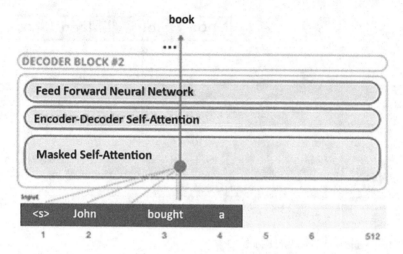

FIGURE 3.6   Attention in a *decoder*-only transformer.

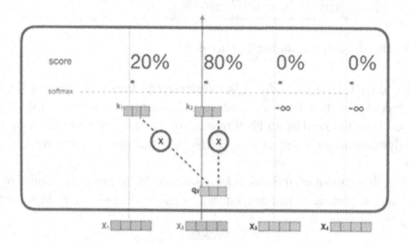

FIGURE 3.7   Masked self-attention in GPT-2.

it focuses on the first word of the sentence or on the previous word. In the case of GPT-2, it can capture 144 attention patterns, through 12 layers, each with 12 independent attention mechanisms (*heads*).

The use of masked self-attention produces an *autoregressive* architecture; that is, the output of the model at time t is used as input at time $t + 1$, which can continuously predict the next *token* in the sequence. Masked self-attention is identical to self-attention, except when it comes to the evaluation of attention scores; for example, assuming that the model has only two *tokens* as input, and we are observing the second *token*. In this case, the last two *tokens* are masked. Then, the model interferes with the scoring step and always evaluates future *tokens* as 0, so the model cannot reach future words (Figure 3.7).

This *masking* is often implemented as an array called an "attention mask"; for example, think of a sequence of four words, "John bought a book." In a language modeling scenario, this sequence is absorbed in four steps, one per word. Since these models work in

	John	bought	a	book
John	1	0	0	0
bought	0.48	0.52	0	0
a	0.31	0.35	0.34	0
Book	0.25	0.26	0.23	0.26

FIGURE 3.8   Example of a masked attention matrix in GPT-2.

"batches," we can assume a batch size of four for this model, which will process the entire sequence as a single batch.

Next, we compute the evaluation score matrices by multiplying a *query* matrix by a *key* matrix. Next, we assign the attention mask triangle, setting the cells we want to mask to *infinity* (inf) or a very large negative number (i.e., –11,000 million). Finally, the SoftMax function is applied to each row of the matrix to produce the actual scores we use for self-attention. This whole procedure is detailed in the previous chapter. Suppose the matrix, after applying SoftMax, is as shown in Figure 3.8.

In simple terms, this matrix of scores means the following:

- When the model processes the first example in the data set (row #1), which contains only one word ("John"), 100% of its attention will be on that word.

- When the model processes the second example in the *dataset* (row #2), which contains the words ("John bought"), when it processes the word "bought," 48% of its attention will be on "John" and 52% of its attention will be on "bought."

- etc.

We can make GPT-2 work exactly like masked self-attention. However, during evaluation, when the model only adds one new word after each iteration, it would be inefficient to recalculate self-attention along the previous paths for *tokens* that have already been processed.

In this case, we process the first *token* (e.g., "John," from the sequence "John bought"), and GPT-2 must maintain the key and value vectors of the *token* "John." Thus, each self-attention layer maintains its respective key and value vectors for that *token*. In the next iteration, when the model processes the word "bought," it does not need to generate the Q, K and V vectors for the *token* "John" as it simply reuses the ones it stored from the first iteration (Figure 3.9).

As a consequence, instead of training a new model for each application, a single model can be pre-trained and then tuned to solve multiple tasks. On the other hand, this approach helps to solve data sparsity problems by pre-training on a large and diverse *dataset*. An efficient approach to adapt a basic model to a specific subsequent task is through *zero- or few-trial type inferences*.

3.3.1.1 Zero/Few-Trials Inference through Prompts

GPT-based models receive text as input and produce text as output. This generic input-output structure can be exploited by providing inputs such as the following:

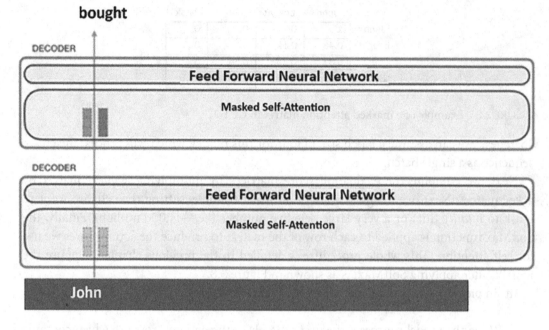

FIGURE 3.9   Masked attention *decoder* blocks for "John bought a book"

```
"Translate this sentence into English: <sentence>=>"
"Summarize the following document: <document>=>."
```

These *prompts* (Wei et al., 2023) for task solving allow zero-shot inference with LM, that is, without seeing examples of the correct output. Given a *prompt*, the most appropriate LM output should solve the task; for example, translating into English or summarizing a document. To perform a few-shot inference, a similar *prompt* can be constructed with examples of correct output provided at the beginning.

In general, there are three types of shots that can be used (without a gradient update):

1. *Zero shot*: the model predicts the answer given only a description of the task; for example:

   ```
   Translate from English to        ← Task description
   Spanish:
   Car =>                           ← Prompt
   ```

2. *One shot*: in addition to the task description, the model sees a simple example of the task; for example:

   ```
   Translate from English to        ← Task description
   Spanish:
   Ice cream => helado              ← Example
   Car =>                           ← Prompt
   ```

3. *Few shots*: in addition to the task description, the model sees a few examples of the task; for example:

```
Translate from English to          ← Task description
Spanish:

Ice cream => helado                ← Examples
Neural nets => redes neuronales    ← Examples
Car =>                             ← Prompt
```

The simplest way to use a pre-trained GPT-2 is to allow it to wander on its own, i.e., generate unconditional samples. Alternatively, a *prompt* can be provided to talk about a specific topic. In the case of rambling, one can simply hand over the start *token* and have it start generating words.

The model has only one input *token*, so that path would be the only active one. The *token* is processed successively through all the layers; then, a vector is produced along that path. That vector can be evaluated against the model's vocabulary (i.e., all the words known to the model: 50,000 words, in the case of GPT-2). In this case, the *token* with the highest probability is selected. However, we can mix things up: we already know that, if we keep clicking on the suggested word in a keyboard application, sometimes you can get stuck in repetitive cycles where the only way out is if you click on the second or third suggested word. The same thing can happen here. GPT-2 has a parameter called top-k, which we can use to make the model consider sampling words other than the main word (e. g., when *top-k* = 1).

In the next step, we add the output of the first step to our input sequence and have the model perform its next prediction. Thus, each layer of GPT-2 retains its own interpretation of the first *token* and uses it to process the second *token*.

GPT uses a 12-layer *decoder*-only *transformer* architecture identical to the original *transformer decoder* (see Figure 3.10). First, GPT performs LM pre-training on a large *dataset* (i.e., BooksCorpus). Then, separate *fine-tuning* is performed in a supervised manner on a variety of discriminative language understanding tasks, such as classification, similarity search, or *entailment*, among others.

Thus, instead of modifying the GPT architecture to solve different tasks, information is provided in a task-specific structure and then the model output is passed to a separate classification layer; for example, in *entailment* tasks, the input sentences are separated and concatenated with a special delimiter. This is provided as input to GPT and then the model output is passed to a separate classification layer. Recall that the textual *entailment* task (aka natural language inference) is a directional relationship between text fragments, so the relationship is fulfilled whenever the truth of one text fragment is derived from another text.

Both GPT and GPT-2 perform pre-training for the purpose of language modeling, but they do not perform *fine-tuning*, but solve the tasks in a *zero-shot* fashion. This means that GPT-2 performs multi-task learning based on two capabilities: pre-training of a generic LM and use of textual *prompts* to perform *zero-shot* inferences.

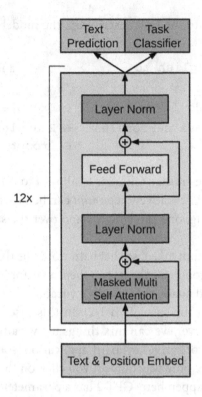

FIGURE 3.10   Architecture of GPT-2.

TABLE 3.1   Hyperparameters of the GPT-2 Architecture for Four Sizes

Parameters	Layers	d_{model}
117M	12	768
345M	24	1024
762M	36	1280
1542M	48	1600

The pre-training is performed on a custom public text *dataset* (i.e., WebText) that is built by extracting popular links from Reddit. The model has been pre-trained in four different sizes as shown in Table 3.1, where d_{model} is the model dimension (i.e., *embedding* size).

In general, the GPT-2 architecture is identical to GPT, except for some minor differences (i.e., different weight initialization, larger vocabulary, longer input sequence, etc.). Despite the size of these LLMs, *underfitting* of the WebText *dataset* usually occurs during pre-training, indicating that LLMs would perform even better.

On the other hand, GPT-2 has been evaluated on several tasks, where it achieves promising results, such as language modeling and reading comprehension, but does not reach the baseline for summary and question-answer generation. However, recall that GPT-2 does not perform *fine-tuning* to solve any of these tasks, so the results are achieved through *zero-shot* inference. Interestingly, *zero-shot* performance improves steadily with LM size, indicating that, by increasing the size of a model, its ability to learn relevant *feature s* during pre-training improves.

3.3.2 The GPT-3 Model

One of the disadvantages of the GPT-2 model is that it is quite limited by the quality and quantity of data used for its pre-training. Thus, if the data are biased or limited, the model may not produce accurate or unbiased results. On the other hand, because the GPT-2 model generates text from the data provided to it, it is possible that it will generate erroneous or inaccurate information. In addition, although the model can produce coherent and well-written text, it usually lacks context. This means that it can be difficult for the model to understand the intent behind a specific question or instruction, which can lead to inappropriate or confusing responses.

In its quest to build more powerful LLMs that do not require *fine-tuning* and *few shots* to understand tasks and perform them, a new version called GPT-3 was developed.

The GPT-3 model has 175 billion parameters (100 times more parameters than GPT-2). Because of this, and the large *dataset* with which it has been trained, it performs well on NLP tasks of zero or few shots; for example, the model can be used to write articles that are difficult to distinguish from those written by humans and can perform tasks that it was never explicitly trained on, such as summarizing numbers, writing *queries*, and generating code in different programming languages, given the task description in natural language, or a *prompt*.

In general, there are two basic concepts introduced by GPT-3:

1. **Learning in context**: LLMs develop pattern recognition and other skills using corpora on which they are pre-trained. As they learn to predict the next word, given context words, these LLMs can recognize patterns in the data that help them minimize the loss of the language modeling task. This capability helps the model during *zero-shot* task transfer. Thus, the model looks for matches in the pattern of examples with what it had learned in the past for similar data and uses that knowledge to perform the tasks.

2. ***Few-shot, one-shot* and *zero-shot* configuration**: these configurations are specialized cases of *zero-shot* task transfer. In the *few-shot* configuration, the model has a description of the task and as many examples as fit the context window. In the *one-shot* configuration, exactly one example is provided to the model, and in the *zero-shot* configuration, no examples are delivered.

GPT-3 was trained with a combination of five different corpora, each of which was assigned a certain weight. Thus, high-quality *datasets* were sampled more frequently, and the model was trained for more than one epoch on them. The *datasets* used include CommonCrawl, WebText2, Books1, Books2 and Wikipedia.

The architecture of GPT-3 is the same as that of GPT-2. However, the main differences are in the following:

1. GPT-3 has 96 layers, each of which has 96 attention heads.

2. The size of word *embeddings* was increased to 12 888.

3. The size of the context window was increased from 1024 for GPT-2 to 2048 *token*s for GPT-3.

4. GPT-3 has 175 billion parameters, compared to 1.5 billion for GPT-2. This difference in size translates into a much greater ability to understand and generate natural language.

5. GPT-3 was trained on a much larger variety of tasks and *datasets* than GPT-2. This includes multilingual *datasets*, which allow GPT-3 to generate text in multiple languages.

6. GPT-3 offers greater control over text generation compared to GPT-2; for example, GPT-3 allows the user to specify the length of generated text, subject matter, tone, and style.

7. GPT-3 has demonstrated abilities to complete much more sophisticated tasks than GPT-2; for example, it can perform language translation, unit conversion, and code completion tasks.

When GPT-3 is asked to learn a new task, its weights do not change. However, the input *prompt* is transformed into complex abstractions that can, by themselves, perform tasks that the actual reference model cannot perform. The *prompt* changes GPT-3 each time, making it an expert on the specific task displayed.

Each time you create a *prompt*, you interact with a different GPT-3 model. If we ask it to tell us a story about "elves and dwarves," its internal form will be very different from if we ask it to calculate "2+2." In addition, after analyzing thousands of poets and poems, we can also enter the name of a poet and GPT-3 can create an original poem based on the author's style. Thus, GPT-3 can replicate the texture, genre, rhythm, vocabulary, and style of the poet's previous works to generate other new works.

Although GPT-3 is an automatic completion model, it can be used for various tasks, including a conversation, converting a sentence to a mathematical expression, generating newspaper articles, designing/creating interface layouts, creating pieces of code, mass marketing via the web, etc. On the other hand, GPT-3 is capable of producing high-quality text; however, it sometimes loses coherence when formulating long sentences and repeats sequences of text over and over again. In addition, GPT-3 does not perform very well on tasks such as *entailment*, and some reading comprehension tasks (Prystawski and Goodman, 2023). Another limitation is that it weights each *token* equally and lacks the notion of task- or goal-oriented *token* prediction. In addition, model inference is complex and costly due to its heavy architecture, and there is less interpretability of the language and results generated by the model.

3.3.3 The GPT-4 Model

An improved version of GPT-3 corresponds to GPT-4, which is a multimodal LLM; that is, it accepts image and text inputs, and generates text outputs (OpenAI, 2023; Nori et al., 2023); for example, it is shown a picture of cooking ingredients, is asked what can be done with them, and can respond with multiple options. GPT-4 also extends the maximum input length compared to previous versions, increasing it to a maximum of 32 768 *tokens* (approximately, fifty pages of text).

GPT-4 was trained on both publicly available data and data from external providers (i.e., government documents or academic papers). Then, *fine-tuning* of the model was performed using RLHF, which incorporates human feedback to help the model solve real-world problems effectively (Madaan et al., 2023). In addition, GPT-4 can have more human-like conversations, provide accurate information on a variety of topics, interact with users in more than twenty languages, write tests, complete assignments, and can even analyze infographics (Sanderson, 2023). Its greater capacity is due to the longer context length: a base version with 8192 *tokens* and a larger version with 32,768 *tokens* (Liu et al., 2023).

3.3.4 Reinforcement Learning from Human Feedback

One of the concerns about the application of LLMs is whether they are aligned with human values in terms of security. In general, LLMs should be designed with three principles in mind:

1. **Usability**: the ability of an LLM to follow instructions, perform tasks, provide feedback, and ask relevant questions to clarify the user's intent when necessary.

2. **Truthfulness**: the ability of an LLM to provide objective and accurate information and to recognize its own uncertainties and limitations.

3. **Harmlessness**: the importance of avoiding toxic, biased, or offensive responses and refusing to assist in dangerous activities.

An LLM is considered *aligned* if he or she can successfully adhere to these general guidelines. However, the diverse nature of these issues may require different strategies and approaches for LLMs to respond appropriately to various requests.

One of the innovations to address these problems simultaneously is based on the use of *reinforcement learning from human feeedback* (RLHF). Suppose we have two LLMs: a *baseline* model and a *preference model*. The role of the baseline model is to determine which action a human being would prefer within a given list of possibilities (i.e., two different responses of the baseline model to a user request). This model could assign a numerical score to each action, effectively ranking them according to human preferences, a concept known as the *reward* model, as shown in Figure 3.11.

Using this *reward model*, the reference model can be iteratively refined by altering its internal text distribution to prioritize human-preferred sequences (as indicated by the reward model). Thus, the reward model serves as a means to introduce a *human preference bias* into the reference model.

Next, the goal is to train a reward model. Although several approaches exist, RLHF specifically leverages human feedback to generate a *dataset* of human preferences. This is then used to learn the *reward function* that represents the desired outcome for a particular task. This *human feedback* can be expressed in several ways:

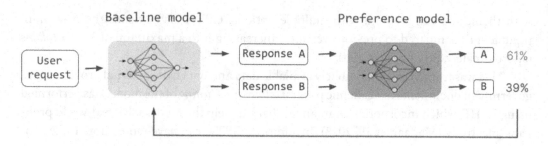

FIGURE 3.11 Preference model for training a reference model.

- **Preference orders**: humans assign a preference order to different outcomes from the reference model.

- **Demonstrations**: instead of evaluating model outcomes, humans perform the full task of writing down preferred responses to a set of prompts.

- **Corrections**: this amounts to editing the output of a model to directly correct undesired behaviors.

- **Natural language input**: instead of directly correcting a model's output, humans must describe a critique of these outputs in natural language.

The *optimal* reward method depends on the specific task to optimize. Once the *reward model* has been established, how is it used to train the reference model? This is where we need techniques such as *Reinforcement Learning* (RL), which focuses on allowing an agent (i.e., an LLM) to learn an *optimal policy* that guides its actions to maximize a *reward*. In this case, the reference model is the agent, and its actions are responses to user input. RL uses the reward model to effectively develop a *human-valued policy* that the LLM will use to generate its responses (Christiano et al., 2023).

Both GPT-3 and GPT-4 use RLHF to integrate pre-training, *fine-tuning*, and *instruction tuning* to improve model generation skills (e.g., conversational capabilities in ChatGPT). In these cases, human feedback is leveraged to produce more engaging, context-sensitive, and safety-aligned responses.

RLHF's overall strategy for *fine-tuning* LLMs consists of three steps:

1. **Collecting human demonstrations**: human annotators are used for a pre-selected set of *prompts*, coming from developers and from model API requests. These demonstrations can be considered as the ideal responses, or responses to these *prompts*, and together they form a training *dataset*. This is then used to perform *fine-tuning* of a previously trained model in a supervised manner, better known as SFT (Supervised Fine-Tuned).

   During SFT, the human demonstration *dataset* is used to fit the reference pre-trained model. Given a *prompt* (P) and an ideal response (A), the base model is asked to calculate the probability that A follows P, which is then used to fit the internal distribution of the model to favor this type of response (see Figure 3.12).

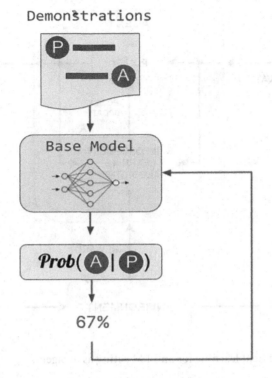

FIGURE 3.12 Supervised fine-tuning process (SFT).

2. **Ordering preferences**: human annotators must vote on one of the SFT model results, thus creating a new *dataset*, composed of comparison data. The reward model is trained on this *dataset*, choosing a list of *prompts*, from which the SFT model generates multiple outputs (between four and nine) for each *prompt*. Annotators rank these outputs from best to worst, forming a new labeled *dataset*, with rankings serving as labels. The comparison data created by the annotators is used to train a reward model that learns to evaluate different responses according to human preferences.

3. **Applying reinforcement learning**: the SFT model teaches the human *preference policy* through the *reward model*. The SFT model is tuned through the reward model, and the result is a policy model. The specific optimization algorithm used to train the policy model is based on the PPO (*Proximal Policy Optimization*) method. This uses a trust region optimization method to train the policy by restricting policy changes within a certain range of the previous policy to ensure stability. This ensures that the policy optimization step does not end up overoptimizing the reward model.

The application of this whole reinforcement learning process for a *rewards* based LLM (agent) and a policy model can be seen in Figure 3.13, where the observations correspond to the ideal *prompts* and responses, and the actions are the ordered preferences respectively, within a fine-tuned *environment*.

Although the collection of demonstrations occurs only once, the second (*reward model*) and third steps (*policy model*) are repeated several times. Thus, more comparison data is

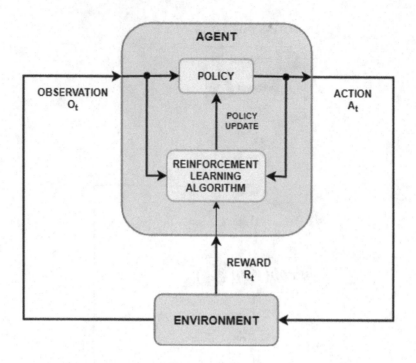

FIGURE 3.13   Reinforcement learning for an LLM acting as an agent.

collected on the current best policy model, which is then used to train a new reward model and, subsequently, a new policy.

The actual effect of performing RLHF *fine-tuning* on an LLM could be seen as follows: the base LLM, trained to approximate the distribution of text extracted from the Internet, possesses one of a chaotic nature, as it has modeled all the text on the Internet, with both extremely valuable and undesirable content.

Suppose we have an ideal base LLM that, at such a stage, is able to perfectly replicate this highly multimodal distribution of text on the Internet; that is, the model has successfully performed the task of finding the perfect distribution. Still, in inference, such an ideal model might exhibit a form of instability in the way it chooses among the millions of modes in the distribution, which together represent the whole cacophony of different tones, sources, and voices that exist in its massive training data.

3.4 PALM

Training larger LLMs is beneficial, but difficult to perform efficiently. Typically, training is distributed across many machines, each with multiple accelerators (i.e., GPU or TPU). This has been done successfully before (i.e., MEGATRON trains a 530 billion-parameter LLM on a system with 2240 GPUs), but the results are not as impressive. However, given higher training throughput, we could (in theory) perform LLM pre-training more extensively on larger *datasets*, which could improve the results.

One way to approach this is to perform pre-training with *cumulative local masks*. This means that the neural network is trained on a secondary task involving the *prediction of hidden word* masks in the text. This is just the goal of the PaLM (Pathways Language Model), which is also based on *transformers* (Chowdhery et al., 2022).

PaLM uses a pre-training technique that involves the accumulation of *local masks*, based on the assumption that words in a text are closely related to words near them (Rohan Anil et al., 2023). This means that, instead of simply *masking* individual words in the text, the model masks groups of words close to each other, allowing for a better understanding of the context and a more accurate representation of the language.

PaLM is a 540 billion-parameter LLM, which eliminates *pipeline* parallelism (Xu et al., 2021), so the architecture achieves higher training performance. This allows PaLM to pre-train on a larger *dataset*, which also enables it to solve difficult reasoning tasks (Yu et al., 2023).

The PaLM model uses a *transformer* only with *decoders*, so it introduces the following changes:

1. **It uses SwiGLU (instead of ReLU) activation functions in the FNN layers**: this corresponds to a combination of Swish and GLU (SwiGLU) activations, which is a product of elements of two linear transforms of the input, to one of which a Swish activation has been applied.

2. **It uses multi-query attention mechanisms in the attention layers**: the multi-head self-attention mechanism is replaced with a *multi-query* attention structure. This only shares *key* and *value* vectors between each of the attention heads, instead of performing a separate projection for each head. This change significantly improves the *autoregressive* decoding efficiency of LLMs. This change is shown in red in Figure 3.14.

3. **It uses parallel transformer blocks**: the *transformer* block is parallelized, instead of the normal (serial) variant, which speeds up the training process by 15% (see Figure 3.15).

4. **It replaces positional embeddings with rotational positional embeddings (RoPE)**: this allows incorporating absolute and relative positioning by encoding the absolute position with a rotation matrix and incorporating the relative position directly into the self-attention. Unlike traditional approaches, where positional vectors are static, RoPE introduces a rotation into these vectors. This allows for capturing the relative relationship between words in different positions, providing richer information about the sequence structure. Rotational positional vectors are computed from a set of basis vectors using a sinusoidal function that varies its frequency and amplitude according to word position.

To understand the impact of the model scale, there are three different sizes of PaLM that have been evaluated, as shown in Table 3.2.

The corpus used for PaLM pre-training consists of 780B *tokens*, extracted from high-quality web pages, books, Wikipedia, news, articles, code and social network conversations.

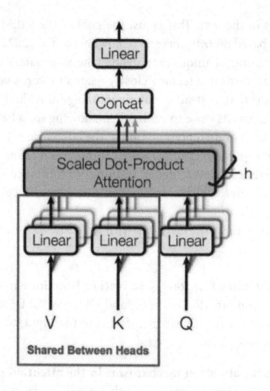

FIGURE 3.14   Multi-query attention mechanism.

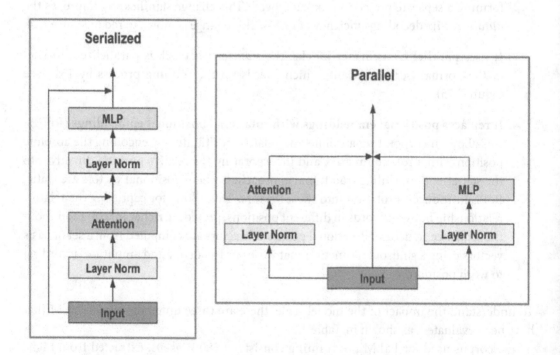

FIGURE 3.15   Serialized (left) versus parallel (right) transformers in PaLM.

TABLE 3.2 Different PaLM Sizes

Size	Layers	Number of Heads	d_{model}
8B	32	16	4096
62B	64	32	8192
540B	118	48	18,432

In addition, this contains 22% of non-English data (see below) and is inspired by corpora used to train other LLMs.

3.4.1 Vocabulary

Since a part of the pre-training corpus is not in English, a *token*izer is incorporated, with a vocabulary size of 256K. The *token*izer takes the raw text input and extracts *token*s (words or subwords) from it. If a *token* is not part of the underlying vocabulary, it is broken down into smaller fragments (i.e., characters), until it has been decomposed into valid *token*s, or replaced with a generic *token* outside the vocabulary.

For multilingual models, typically the size of the underlying vocabulary is greatly increased to avoid this effect, as multi-language data will use a wider range of *token*s. Therefore, PaLM adopts a larger vocabulary size than usual to avoid incorrectly *token*izing the data and allow for more effective learning in multiple languages.

3.4.2 Training

PaLM is trained on a set of 6144 TPU chips that are distributed in two TPU modules, in order to make the model components run in parallel.

Communication is very fast between TPUs, although communication between modules (*pods*) is much slower. This is critical, as parallelizing data and models requires bandwidths that are too large. This is usually addressed using two strategies:

1. Limiting training to a single TPU pod.

2. Using pipeline *parallelism*, which has lower bandwidth requirements between pods.

PaLM is efficiently trained on TPU pods (i.e., a set of TPU devices connected via dedicated high-speed network interfaces) with a combination of model and data parallelism, but without *pipeline* parallelism (i.e., a technique used in data processing and distributed computing to improve the efficiency of processing systems). This allows the processing load to be distributed among several TPUs.

PaLM performs well in many tasks. However, it has some difficulties in solving basic reasoning tasks. This can usually be solved by using *reasoning chain prompts* (i.e., several reasoning steps before the final result) to improve reasoning capabilities.

3.4.3 PaLM-2

An improved version called PaLM-2 incorporates new multilingual, reasoning, and programming capabilities, such as the following:

- **Multilingualism**: the model has been pre-trained on multilingual texts in over a hundred languages. This allows it to improve its ability to understand, generate and translate nuanced text, including idioms, poems, and riddles, in a wide variety of languages.

- **Reasoning**: The PaLM-2 training *dataset* includes scientific articles and web pages containing mathematical expressions. As a result, the model demonstrates enhanced capabilities in logic, common sense reasoning and mathematics.

- **Coding**: the model is pre-trained on a large number of publicly available source code *datasets*. This means that it performs well in popular programming languages such as Python and JavaScript, but also in code in specialized languages such as Prolog, Fortran, and Verilog.

- **Applications**: More than twenty-five Google products and *feature s* are powered. Its enhanced multilingual capability allows it to accommodate new languages. In addition, workspace *feature s* are leveraged to help you write in Gmail and Google Docs, as well as organize spreadsheets, and its healthcare applications, such as Med-PaLM-2, trained by medical teams, can answer questions, and summarize ideas, within a variety of especially dense medical texts.

- **Security**: has a specialized version of Palm-2 called Sec-PaLM, trained in security use cases and cybersecurity analysis. This uses AI techniques to help analyze and explain the behavior of potentially malicious scripts and better detect which scripts are actually threats to people and organizations in unprecedented time.

PaLM-2 outperforms in advanced reasoning tasks, including code and math, classification and question answering, machine translation and multilingual proficiency, as well as natural language generation better than the previous PaLM version. The model is based on an approach to building and implementing AI responsively. To this end, PaLM-2 was rigorously evaluated for its potential harms and biases, capabilities, and downstream uses in research and product applications.

Unlike PaLM, the PaLM-2 model unifies three distinct LLM research advances:

1. **Use of optimal computational scaling**: it scales the size of the model and training *dataset* in proportion to each other. This allows PaLM-2 to be smaller than PaLM, but more efficient with better overall performance, including faster inference, fewer parameters, and a lower service cost.

2. **Improved *dataset* mix**: the model enhances its corpus with a more diverse and multilingual pre-training mix, including hundreds of human and programming languages, mathematical equations, scientific articles, and web pages.

3. **Updated architecture**: PaLM-2 has improved architecture and was trained on a variety of different tasks, all of which help the model learn different aspects of the language.

TABLE 3.3   Differences between PaLM and PaLM-2

Characteristic	PaLM	PaLM-2
Size	540 trillion	1.1 quintillion
Training data	600 trillion	1.65 quintillion
Tasks	Various tasks including translation, encoding, and question answering	Same as previous plus generating creative text, formatting, or music

In order to achieve a responsible and secure LLM, PaLM-2 incorporates three innovations:

1. **Pre-training *datasets***: eliminating forms of sensitive personally identifiable information, filtering duplicate documents to reduce memorization, and sharing analysis of how people are represented in pre-training data.

2. **New capabilities**: the model shows improved multilingual toxicity classification capabilities and has built-in control over toxic generation.

3. **Evaluations**: potential harms and biases are evaluated in a variety of potential downstream uses of PaLM-2, including dialogue, classification, translation, and question answering. This includes the development of new assessments to measure potential harms in generative question-answering and dialog environments related to the harms of toxic language and social bias related to identity terms.

Finally, in Table 3.3, the main differences between PaLM and PaLM-2 are shown.

3.5 LLaMA

With recent advances in LLM, full access to research is still limited to the resources required to train and run such large models. This restricted access has limited the ability to understand how and why these models work, hindering progress in efforts to improve their robustness and mitigate known problems such as bias, toxicity, and the potential to generate misinformation. Alternatively, smaller models trained on more *tokens* are easier to retrain and tune for specific potential product use cases.

To address these problems, LLaMA (Large Language Model Meta AI) arises to aid research and development in NLP (Guilla, 2023). The model is smaller and higher performing compared to other LLMs, allowing others in the research community who do not have access to large amounts of infrastructure to study these models (Touvron et al., 2023). Training the smaller model is desirable because it requires much less computational power and resources to test new approaches (Zhang et al., 2023). Basic models are trained on a large unlabeled *dataset*, which makes them ideal for performing *fine-tuning* on various tasks. Currently, LLaMA is available in various sizes and configurations, as shown in Table 3.4.

The model was trained with texts from the 20 most widely spoken languages, focusing on those with Latin and Cyrillic alphabets. As a basic model, LLaMA is designed to be versatile and can be applied to many different use cases, as compared to a tuned model that is designed for a specific task.

TABLE 3.4   Hyperparameters of Pre-Trained LLaMA Models

Size	d_{model}	Number of Heads	Number of *Tokens*
7B	4096	32	1T
13B	5120	40	1T
33B	6656	52	1.4T
65B	8192	64	1.4T

3.5.1 Pre-Training Data

The training *dataset* for LLaMA is a mixture of several publicly accessible and open source compatible data sources. In total, this source accumulates 1452 TB and includes:

- **English CommonCrawl** (67%): five sources (2017 through 2020) were included, which were processed for language identification with a linear classifier, in order to remove non-English languages and filter out low-quality content with an *n-gram* model.

- **C4** (15%): this was preprocessed for language identification and filtered using heuristics such as the presence of punctuation marks or the number of words and sentences in a web page.

- **Github** (4.5%): low-quality files were filtered with heuristics based on line length or alphanumeric character ratio, removing headers.

- **Wikipedia** (4.5%): data covering 20 languages (period June–August 2022) were included and processed to remove hyperlinks, comments, and other formatting data.

- **Gutenberg and Books3** (4.5%): public domain books and the *Books3* section of *ThePile* were included.

- **ArXiv** (2.5%): latex files from *arXiv* were included and processed to add scientific data to the *dataset*.

- **Stack Exchange** (2%): this high-quality question and answer website, covering a diverse set of domains from computer science to chemistry, was included and data from the 28 largest websites were maintained.

This allowed the creation of a training *dataset* containing approximately 1.4T *tokens*.

3.5.2 Architecture

The main architecture of LLaMA is based on *transformers*, which incorporate three relevant changes:

1. **Prenormalization**: it is based on GPT-3 and allows improving training stability by normalizing the input of each *transformer* sublayer, using the RMSNorm normalization method (Phuong and Hutter, 2022).

TABLE 3.5 Hyperparameters of the Different LLaMA Configurations

Size	d_{model}	Number of Heads	Number of Layers	Number of *Tokens*
6.7B	4096	32	32	1T
13B	5120	40	40	1T
32.5B	6656	52	60	1.4T
65.2B	8192	64	80	1.4T

2. **SwiGLU activation function**: it is based on PaLM and replaces the nonlinearity of *ReLU* with the *SwiGLU* activation function (Ekman, 2022).

3. **Rotational *embedding***: it is based on the GPT model and adds RoPE-type *embeddings*, at each layer of the network.

The main hyperparameters of the different variations of the model are shown in Table 3.5.

In addition, several optimizations were performed to improve the training speed of the models:

1. **Efficient implementation of causal multi-headed attention**: this allows to reduced memory usage and execution time. This is achieved by not storing attention weights and not calculating key/*query* scores that are masked, due to the causal nature of the language modeling task.

2. **Reduced number of activations that are recalculated during the backward pass with checkpoints**: activations that are costly to compute, such as linear layer outputs, are stored by manually implementing the inverse function for the *transformer* layers.

3. **Reduced model memory usage**: parallelism of models and sequences is used. In addition, the computation of activations and inter-GPU communication over the network was superimposed.

In the final model training results, several NLP benchmark tasks and learning styles, such as *zero shot* and *few shots*, were considered.

Overall, the model is available in two versions: LLaMA-7B and LLaMA-13B, and LLaMA-33B and LLaMA-65B, of 7B, 13B, 33B, and 65B parameters, respectively. In addition, variations are available such as Alpaca (7B and 13B, trained on GPT-3 instructions), and Vicuna (7B and 13B, trained on ChatGPT conversations) (Darling, 2022).

3.6 LANGUAGE MODEL FOR DIALOGUE APPLICATIONS (LaMDA)

LaMDA (Language Model for Dialogue Applications) is a family of LLMs for conversational systems applications developed by Google (Thoppilan et al., 2022). LaMDA has gone through several improvements and the current generation is based on *transformers*, which only uses the *decoder* module. The model has been pre-trained on a text corpus including both documents and dialogs consisting of 1.56 billion words and then trained with specific data by *fine-tuning*, generated from manually annotated responses that meet sensitivity and

TABLE 3.6   Hyperparameters of Different Configurations of LaMDA

Size	Number of Layers	d_{model}	Number of Heads
2B	10	2560	40
8B	16	4096	64
137B	64	8192	128

safety criteria. The main *feature* is a mechanism in which the *transformer* interacts with an external information retrieval system to improve the accuracy of the responses generated to a user. The main hyperparameters of the model versions are shown in Table 3.6.

Unlike other models, LaMDA was trained on dialog data. Because of this, during its training, the model captures several of the nuances that distinguish an open conversation from other forms of language. One such nuance is reasonableness, which basically answers "does the response make sense to a given conversational context?"; for example, if someone says:

```
I just started a new job.
```

One might expect another person to respond with something like:

```
How exciting: a friend got a job with an excellent work
environment.
```

This response makes sense, given the initial expression. However, *sensibleness* is not the only thing that makes a good response. If so, the sentence "that's fine" is a sensible response to almost any statement, just as "I don't know" is a sensible response to most questions. Satisfactory answers also tend to be specific, relating clearly to the context of the conversation. In the example above, the answer is sensible and specific.

LaMDA is based on a previous model called Meena, which was fully trained with 2.6 billion parameters. Meena can conduct conversations that are more sensible and specific than those of existing state-of-the-art *chatbots*. This improvement is made possible by using a metric called Sensibleness and Specificity Average (SSA), which captures basic attributes for human conversation. Interestingly, the metric is strongly correlated with other standard measures, such as perplexity. Once trained and validated with these metrics, LaMDA can be adjusted to significantly improve the reasonableness and specificity of its responses.

3.6.1 Objectives and Metrics

The results generated by LaMDA are mainly evaluated according to three criteria:

1. **Quality**: this is usually decomposed into three dimensions – Sensibleness, Specificity, and Interestingness (SSI) – which are measured by human evaluators. Sensibleness refers to whether the model produces answers that make sense in the context of the dialogue (i.e., no common sense errors, no absurd answers, and no contradictions). Specificity is measured by judging whether the system's response is specific to the dialogue context above and not a generic response that could apply to most contexts (i.e.,

"okay" or "don't know"). Finally, Interestingness ("interest") measures whether the model produces responses that are also insightful, unexpected, or witty and therefore more likely to create a better dialogue.

2. **Safety**: this is composed of a set of safety objectives that capture the behavior that the model should exhibit in a dialogue. These objectives attempt to constrain the model's output to avoid any undesirable outcomes that create risks of harm to the user and to avoid reinforcing unfair bias; for example, these objectives train the model to avoid producing output that has violent or gory content, promotes slurs or hateful stereotypes toward groups of people, or contains profanity.

3. **Grounding**: this is defined as the proportion of responses with statements about the external world that can be supported by authoritative external sources, as a share of all responses containing statements about the external world. A related metric, Informativeness, is defined as the proportion of responses with information about the external world that can be supported by known sources, as a share of all responses. Thus, casual responses that do not carry any real-world information (i.e., "That's a great idea") affect Informativeness, but not *Grounding*.

3.6.2 Pre-Training of LaMDA

Pre-training of the model is performed in two stages: pre-training, through *fine-tuning*, and *grounding*. In the pre-training stage, a *dataset* of 1.56T words, almost forty times more words than those used to train previous dialog models such as Meena, is created from public dialog data and other web documents. After *tokenizing* the *dataset* into 2.81T *tokens*, model pre-training is performed using techniques for parallelizing neural networks such as GSPMD, in order to predict the next *token* in a sentence, given the previous *tokens*.

- **Fine-tuning**: the model is trained to perform generative tasks that produce responses to given contexts, as well as classification tasks, on whether a response is safe and of high quality. This produces a single multitasking model that can do both. The generator is trained to predict the next *token* over a dialogue *dataset* restricted to the back-and-forth dialogue between two authors. On the other hand, the classifiers are trained to *predict safety and quality assessments (aka SSI)*, for in-context response using annotated data. During a dialog, the generator produces several candidate responses, given the current multi-turn dialog context, and the classifiers predict the safety and SSI assessments for each candidate response. First, candidate responses whose evaluations or safety scores (i.e., SSI) are low are filtered out. The other candidates are re-ranked according to their SSI scores and the one with the highest score is chosen as the answer (see the example in Figure 3.16). All this allows for an increase in the density of high-quality candidate responses.

- **Grounding**: While people are able to verify their facts by using tools and referring to established knowledge bases, many LLMs draw their knowledge only from the parameters of their internal model. To improve the robustness of the original

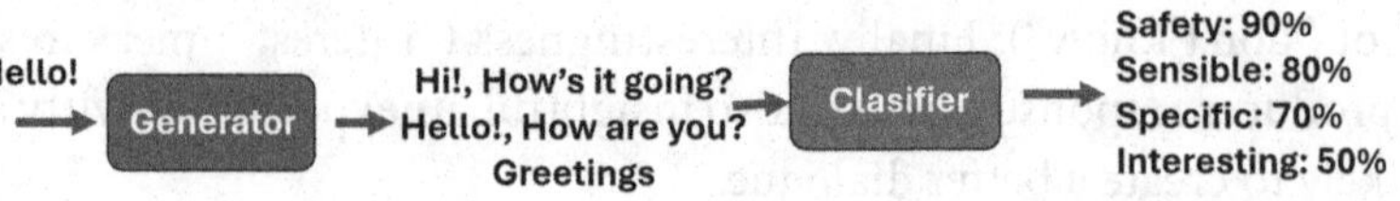

FIGURE 3.16   Evaluation of candidate answers according to safety and interest.

LaMDA response, a *dataset* of dialogues between people and LaMDA was collected and annotated with information retrieval *queries* and results obtained, where appropriate. Then, the generator and classifier are tuned on this *dataset* to learn how to invoke an external information retrieval system during its interaction with the user, to improve the *grounding* of its responses.

3.7 MEGATRON

Megatron is an LLM developed by the NVIDIA company and is one of the world's largest and most powerful models of its kind, which also uses a *transformer* architecture. The model has been trained with a large amount of text data to learn to predict the next word or phrase in a text sequence. The model has billions of parameters and can generate consistent, high-quality text in several languages, including English, Spanish, French, German, Italian and many more.

Megatron is composed of several parts and technologies that work together to create the model, the main components of which include:

- **Transformer**: this architecture is used to process variable-length text sequences and capture long-term relationships between words.

- **Large-scale parallelization**: it uses large-scale parallelization techniques to train the LLM on multiple GPUs. This allows the model to process large amounts of text data and significantly accelerates the training process.

- **Memory optimization**: it uses memory optimization techniques to reduce the amount of memory required to train the model. This allows the model to be trained on GPUs with less memory, reducing hardware costs.

- **Precision mixing**: it uses precision mixing techniques to reduce the amount of memory required to store the model parameters. This allows the model to have more parameters and therefore be more accurate without significantly increasing the memory required.

- **Unsupervised learning**: the model is trained using unsupervised learning techniques, which means that the model learns from large amounts of text data without the need for annotations or labels. This allows the model to be trained on any language or text domain without the need for labeled data.

Unlike other LLMs, Megatron takes advantage of the parallel nature of the *transformer* structure and is able to optimize it for large numbers of parameters. Thus, it performs a parallel implementation of a simple model by adding some synchronization primitives. In particular, Megatron uses two parallelization techniques:

1. **Distributed model parallelism**: the model is split horizontally into several equal parts, each running on a different GPU. This allows it to scale to large GPU systems, resulting in higher training speed and faster throughput; for example, for the self-attention block, the inherent parallelism in the multi-head attention operation associated with the *key (K), query (Q), and value (V)* in a parallel column is exploited, such that the multiplied matrix corresponds to each attention head running locally on a GPU. This allows splitting the attention and workload parameters across GPUs and does not require any immediate communication to complete the self-attention. The linear layer is then parallelized along its rows and takes the output of the parallel attention layer directly, without the need for communication between GPUs.

2. **Distributed data parallelism**: the input data is divided into several chunks and distributed among the GPUs. Each GPU processes its own *dataset* and then combines the results to produce the final output. In particular, recall that a *transformer* has output *embeddings* with the dimension of hidden size (*H*) by vocabulary size (*v*). Since the vocabulary size is on the order of tens of thousands of *tokens* for LLM, it is very beneficial to parallelize the computation of the output *embeddings*.

Much of the parallelism approach is based on techniques aimed at reducing communication and maintaining GPU computational limits. Instead of having one GPU compute part of the *dropout*, normalizing layers, or residual connections, and passing the results to other GPUs, we simply duplicate the computation between GPUs, as shown in Figure 3.17.

For this, duplicate copies of the layer normalization parameters are maintained on each GPU. Then, the output of the model parallel region is taken, and dropout and residual connections are run on these tensors before passing them as input to the following parallel regions. To optimize the model, each parallel worker process optimizes its own parameters, which avoids the need for inter-GPU communication. The different configurations of the model can be seen in Table 3.7.

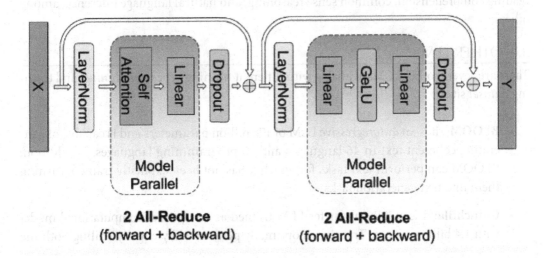

FIGURE 3.17   Model parallelism in Megatron.

TABLE 3.7   Hyperparameters of the Megatron Model

d_{mode}	Number of Heads	Number of Layers	Size	GPU for Model Parallelism	GPU for Model and Data Parallelism
1536	16	40	1.2 B	1	64
1920	20	54	2.5 B	2	128
2304	24	64	4.2 B	4	256
3072	32	72	8.3 B	8	512

In general, the *transformer* structure was reorganized to allow this parallelization, changing the order of the normalization layer and the residual connections, allowing to scale the model with less training *loss*.

3.7.1 Training Data

To collect diverse training *datasets* with long-term dependencies, larger *datasets* were created, including Wikipedia, CC-Stories, RealNews, BookCorpus, and OpenWebText. In addition, unnecessary information was removed from some corpora. These *datasets* were combined and documents with content smaller than 128 *tokens* were filtered out. The training used the *transformer decoder* architecture, with 530 billion parameters, the total number of layers, hidden dimensions, and attention heads being 105, 20,480, and 128, respectively.

Some improvements to the original model made possible the creation of Megatron-Turing, which uses NVIDIA's Turing processor architecture, with 530 billion parameters. The model is designed to take full advantage of the processing power of Turing-based GPUs, resulting in increased speed and efficiency in LLM training.

This power optimizes matrix computations and the implementation of a low-latency communication model between GPU nodes. In addition, Megatron-Turing uses a low-precision model compression scheme, which reduces the size of the model and accelerates its processing on the GPU. By combining the Turing processor architecture with Megatron's parallelization technique, billion-parameter LLMs can be trained in a reasonable time, and with good accuracy on various natural language tasks such as completion prediction, reading comprehension, common sense reasoning, and natural language inference, among others.

3.8 OTHER LLMS

There are several other LLMs available with different training types, *fine-tuning*, and training *dataset* sizes[2]:

- **BLOOM**[3]: it is an *autoregressive* LLM of 176 billion parameters and is capable of generating coherent text in 46 languages and 13 programming languages. In addition, BLOOM can perform text tasks for which it has not been explicitly trained, turning them into text generation tasks.

- **Chinchilla**: it is a 70B-parameter LLM trained as an optimal computational model with 1.4 billion *tokens*. Training is optimally performed by equally scaling both the size of the model and the number of training *tokens*: the model uses four times more training data than other models.

- **AlexaTM**[4] (Alexa Teacher Model): it is a 20B-parameter LLM based on *one-shot* and *few-shot* tasks. The model outperforms GPT-3 on several benchmark tasks and has less than 1/8 the number of parameters. Unlike other models, AlexaTM uses both the *encoder* and the *transformer decoder*. The *encoder* stage enables better performance in summary generation and machine translation tasks compared to *decoder*-only LLMs.

- **Dolly**[5]: it is an open source LLM that has been pre-trained for instruction tracking, with 6 billion parameters, compared to 175 billion for GPT-3. Dolly is a clone of another open source model called Alpaca, inspired by LLaMA.

- **Alpaca**[6]: it is an improved LLaMA LLM with seven billion parameters for instruction tracking. Alpaca behaves qualitatively similar to GPT-3 but is smaller and cheaper to reproduce. The model was trained with 52K demos by *fine-tuning*.

- **Falcon**[7]: it is an *autoregressive decoder* LLM, with 40B parameters trained on a billion *tokens*. The model uses only 75% of the training computation from GPT-3, 40% from Chinchilla and 80% from PaLM-62B. Falcon can extract high-quality content from web data and use it to train a custom code base. Falcon's architecture was optimized for performance and efficiency, outperforming GPT-3 by only 75% of the training computation budget and requiring one-fifth of the computation at inference time. One *feature* of the model is the quality of its training data, which is based predominantly (over 80%) on RefinedWeb, a web *dataset* based on CommonCrawl. A smaller version, Falcon-7B, is a 7B parameter causal *decoder*-only model trained on 1500B *tokens* from RefinedWeb.

- **ORCA**: is a Microsoft-developed LLM with 13B parameters which learns from GPT-4 *prompts*, including explanation traces, step-by-step thought processes and other complex instructions. The main difference between Orca 13B and GPT-4 is the use of explanation traces; for example, it allows ORCA to understand the underlying logic behind the responses generated by GPT-4. This allows it to understand the context and subtleties of different scenarios, significantly improving reasoning and comprehension skills.

- **Mistral**[8] is a 7-billion-parameter language model released by Mistral AI. The model provides both efficiency and high performance to enable real-world applications. Due to its efficiency improvements, the model is suitable for real-time applications where quick responses are essential. Mistral 7B outperformed the best open source 13B model (Llama 2) in all evaluated benchmarks. In addition, the model uses two attention mechanisms: (1) Grouped-query attention for faster inference and reduced memory requirements during decoding, and (2) Sliding window attention for handling sequences of arbitrary length with a reduced inference cost.

3.9 CONCLUSIONS

The advancement of LLMs has opened up new possibilities and opportunities for industries and society. From *chatbots* to language translation, LLMs have the potential to revolutionize the way we communicate and interact with technology. A key *feature* of these models is their ability to efficiently represent the contexts of words within a sentence. However, this requires a large amount of data to be effective; hence, few-shot learning techniques, which aim to allow models to learn from a limited amount of data, have the potential to greatly improve the flexibility and efficiency of models.

On the other hand, current LLMs require large amounts of labeled data to process language accurately. However, unsupervised learning techniques, which do not require labeled data, have the potential to greatly reduce the amount of data needed for NLP models to be effective.

NOTES

1 https://www.aidemos.info/
2 https://crfm.stanford.edu/helm/latest/
3 https://huggingface.co/bigscience/bloom
4 https://github.com/amazon-science/alexa-teacher models
5 https://github.com/databrickslabs/dolly/
6 https://github.com/tatsu-lab/stanford_alpaca
7 https://falconllm.tii.ae/
8 https://huggingface.co/docs/transformers/main/en/model_doc/mistral

## CHAPTER 4

Model Evaluation

4.1 INTRODUCTION

The development and evaluation of LLMs has become a fundamental area of research in the field of NLP. These models are trained on vast amounts of data and have the ability to generate consistent, quality text. However, evaluating the quality and performance of these models is a challenge in itself (Stephanie Lin, 2022).

First, evaluation of LLMs requires understanding the main tasks on which LLMs can be measured, because the complexity of a text generation task and a classification or reasoning task have completely different complexities.

On the other hand, the evaluation of LLMs involves the selection of appropriate metrics. Although traditional metrics such as perplexity and grammatical accuracy are useful, they do not fully capture the quality and consistency of the text generated by these models. Therefore, new metrics, such as lexical diversity and human evaluation, are being explored that consider broader aspects of text generation.

In addition, the creation of suitable evaluation corpora or *datasets* is a major challenge. High-quality benchmark data is essential for comparing model performance. These *datasets* must be representative and cover a wide range of domains and language styles.

Thus, LLM evaluation is an active area of research seeking to develop more robust metrics and appropriate evaluation *datasets*. The goal is to accurately measure the quality and performance of these models in text generation, overcoming the limitations of traditional metrics and addressing the challenges of contextualized assessment. These rigorous evaluations are essential to continuously improve and refine these models, and to ensure that their application in various areas is reliable and effective.

4.2 EVALUATION TASKS

In general, LLMs are evaluated on a diverse set of benchmark NLP tasks, such as:

- **Language modeling**: the performance of a language model can be evaluated using the word prediction task. This involves presenting the model with a sequence of words and assessing its ability to predict the next word.

DOI: 10.1201/9781003517245-4

- **Text generation**: the model's ability to generate coherent and meaningful text can be assessed using the sentence or paragraph completion task. This involves presenting the model with an incomplete sentence or paragraph and assessing its ability to complete it in a coherent and meaningful way.

- **Machine translation**: the model's ability to automatically translate text from one language to another can be assessed using the machine translation task. This involves presenting the model with a text in a source language and assessing its ability to translate it into the target language.

- **Question answering**: the model's ability to answer questions based on a given context can be assessed using the question-answering (QA) task. This involves presenting the model with a question and a related context and assessing its ability to provide a relevant and accurate answer.

- **Natural language understanding**: the model's ability to understand natural language can be assessed using the semantic analysis task. This involves presenting the model with a text and assessing its ability to understand its structure and meaning.

Suppose two LLMs are trained for a simple question-answer system, one with attention and one without attention, or one with more layers of processing than the other. How can these models be evaluated to find the one best suited to the task? Quite simply, through *benchmarking*.

To examine the effectiveness of LLMs, there are several tasks and benchmarks for conducting an empirical evaluation and analysis. First, three basic types of LLM evaluation tasks for language generation and comprehension are introduced; then, advanced LLM tasks with more complex settings or goals are described.

4.2.1 Basic Evaluation Tasks

This group mainly includes three types of LLM evaluation tasks:

a. **Language generation**: this type of task can be classified into three groups:

- **Language modeling**: this aims to predict the next *token* based on the previous *token*s, which mainly focuses on basic language comprehension and generation ability.

- **Conditional text generation**: this task focuses on generating text that satisfies specific task demands based on the given conditions, typically including machine translation, text summarization, and question answering.

- **Code synthesis**: existing LLMs show strong skills in generating formal language, especially computer programs that satisfy specific conditions, called "code synthesis." Since the generated code can be directly verified by execution with the corresponding compilers or interpreters, existing work mainly evaluates the quality of code generated from LLMs by calculating the pass rate in test cases.

b. **Knowledge utilization**: this is an important skill of systems that perform knowledge-intensive tasks (e.g., answering common sense questions and completing facts) based on factual evidence. This requires LLMs to appropriately use factual knowledge from the pre-training corpus or retrieve external data when necessary. In particular, Question-Answering or QA and knowledge completion have been two commonly used tasks to assess this skill. According to the testing tasks and assessment settings (with or without external resources), knowledge utilization tasks are classified into three types:

- **Closed-book QA**. These test the factual knowledge acquired from the LLMs of the pre-training corpus, where models must answer the question only based on the given context, without using external resources. Performance on this type of test shows a scaling law pattern in terms of model and data size: scaling the training parameters and *tokens* can increase the capacity of LLMs and help them learn (or memorize) more knowledge from the pre-training data.

- **Open-book QA**. In this task, useful evidence can be extracted from an external knowledge base or a set of documents and then answered based on the extracted evidence. To select relevant knowledge from external resources, LLMs are combined with a text retriever (or even a search engine), which is trained independently or jointly with LLMs.

- **Knowledge completion**. LLMs can be considered a knowledge base that can be leveraged to complete or predict missing parts of knowledge units (e.g., knowledge triples). Such tasks can assess how much and what kind of knowledge LLMs have learned from the pre-training data.

c. **Complex reasoning**: this task refers to the ability to understand and use supporting evidence or logic to derive conclusions or make decisions. According to the type of logic and evidence involved in the reasoning process, assessment tasks can be divided into three main categories:

- **Knowledge reasoning**: this task relies on logical relationships and evidence about factual knowledge to answer a given question. Usually, specific *datasets* are mainly used to assess the reasoning ability of the corresponding type of knowledge. In addition, the quality of the generated reasoning process is evaluated, through automatic metrics or human evaluation. Generally, these tasks require LLMs to perform "step-by-step" reasoning based on factual knowledge, until the answer to the given question is reached (Wei et al., 2023). To obtain step-by-step reasoning ability, a strategy known as *chain of thought* (i.e., "chain of thought" or CoT) is usually applied to improve complex reasoning ability. CoT involves intermediate reasoning steps, which can be manually created or automatically generated in *prompts*, to guide the model to perform multistep reasoning (Wang, 2023).

- **Symbolic reasoning**: these types of tasks focus primarily on the manipulation of symbols in a formal rule setting to accomplish some specific goal, where the operations and rules may never have been seen by pre-trained LLMs.

- **Mathematical reasoning**: this uses mathematical knowledge, logic and computation to solve problems or generate proof statements. Existing mathematical reasoning tasks can be mainly classified into mathematical problem solving and automated theorem proving. Since these tasks also require multistep reasoning, the CoT strategy has been adopted to improve reasoning. Since mathematical problems in different languages share the same mathematical logic, a benchmark of multilingual mathematical verbal problems can also be used to assess the multilingual mathematical reasoning ability of LLMs.

4.2.2 Advanced Assessment Tasks

LLMs also exhibit some superior skills that require special considerations for assessment, such as the following:

a. **Human alignment**:

   LLMs are expected to be able to align with human values and needs (Wang et al., 2022). To assess this capability, multiple human alignment criteria, such as kindness, honesty, and safety, are considered. For kindness and honesty, adversarial question-answering tasks (i.e., TruthfulQA) can be used to examine LLMs' ability to detect possible falsehoods in a text. In addition, innocuousness (i.e., harmlessness) can also be assessed by several existing benchmarks. However, human evaluation remains a more direct way to effectively test LLMs' human alignment capabilities.

b. **Interaction with external environment**:

   LLMs have the ability to receive feedback from the external environment and perform actions according to behavioral instructions, such as generating natural language action plans to manipulate agents (Bernstein, 2023). This capability also arises in LLMs that can generate detailed and very realistic action plans, whereas smaller models tend to generate shorter or meaningless plans.

c. **Tool manipulation**:

   When solving complex problems, LLMs can use tools if they determine it necessary; for example, through APIs. Thus, the web browser plug-in allows ChatGPT to access up-to-date information. In addition, the incorporation of third-party plug-ins is particularly key to creating a thriving ecosystem of LLM-based applications. To assess this capability, complex reasoning tasks are usually adopted for evaluation, such as mathematical problem solving or open-book QA, where successful use of tools is very important to improve LLM skills.

4.2.3 Regulatory Compliance Tasks

LLMs and other generative AI applications are transforming society with their capabilities. However, there are varied issues related to their security, transparency, data usage, privacy, and bias, among others, that need to be evaluated according to their level of *compliance* (i.e., compliance) with certain international regulations (Liang et al., 2021). Recently, the

European Union (EU) finalized its AI law as the world's first comprehensive regulation to regulate AI. The European Parliament adopted a draft of the law that includes explicit obligations for basic model providers.

The Stanford Institute for Human-Centered Artificial Intelligence[1] identified and synthesized the requirements of the AI law adopted by the EU, which can be seen summarized in Table 4.1. With this approach, the following information was extracted to evaluate different models and the reference documents:

- Almost 22 requirements addressed to foundation model providers were extracted from the EU version of the law. Twelve of the 22 requirements were selected for evaluation.

- The 12 requirements were categorized in relation to (*i*) data resources (3), (*ii*) IT resources (2), (*iii*) the model itself (4), or (*iv*) implementation practices (3). Many of these requirements focus on transparency, for example, disclosure of what data were used to train the basic model, how the model performs on standard benchmarks, and where it is implemented.

- A 5-point rubric was designed for each of the 12 requirements. While the law sets out high-level obligations for core model providers, it is not clear how these obligations are to be interpreted or enforced. These rubrics can directly inform legal interpretation or standards, even in areas where the language of the law is particularly confusing.

- We evaluated 10 model language providers' compliance with 12 of the law's requirements for foundation models under our rubrics. The vendors were evaluated independently among several human evaluators, with substantial inter-rater agreement of Cohen's kappa = 0.74.

The final evaluation is shown in Figure 4.1, where both LLMs and other types of generative models for images are shown.

The results show a surprising range in compliance among the vendors: some vendors score below 25% (i.e., AI21 Labs, Aleph Alpha, Anthropic...) and only one vendor scores at least 75% (Hugging Face/BigScience). Even for the vendors with the highest score, there is still significant room for improvement. This confirms that the law (if enacted) would bring about significant change in the ecosystem, making substantial progress toward greater transparency and accountability.

There are four areas in which many organizations receive low scores (generally, 0 or 1 out of 4): (1) copyrighted data, (2) computation/power, (3) risk mitigation, and (4) evaluation/testing. These speak to established issues in the scientific literature:

- **Unclear liability due to copyright**: Few vendors disclose information on the copyright status of training data. Many basic models are trained on data culled from the Internet, a substantial fraction of which is likely to be copyrighted.

TABLE 4.1   Regulations and Compliance Proposed by the European Union

Category	Keyword	Requirement (summarized)	Section
Data	Data sources	Describe data sources used to train the foundation model	Amendment 771, Annex VIII, Section C, page 348
	Data governance	Use data that is subject to data governance measures (suitability, bias, and appropriate mitigation) to train the foundation model	Amendment 399, Article 28b, page 200
	Copyrighted data	Summarize copyrighted data used to train the foundation model	Amendment 399, Article 28b, page 200
Compute	Compute	Disclose compute (model size, computer power, training time) used to train the foundation model	Amendment 771, Annex VIII, Section C, page 348
	Energy	Measure energy consumption and take steps to reduce energy use in training the foundation model	Amendment 399, Article 28b, page 200
Model	Capabilities/limitations	Describe capabilities and limitations of the foundation model	Amendment 771, Annex VIII, Section C, page 348
	Risks/mitigations	Describe foreseeable risks, associated mitigations, and justify any non- mitigated risks of the foundation model	Amendment 771, Annex VIII, Section C, page 348 and Amendment 399, Article 28b, page 200
	Evaluations	Benchmark the foundation model on public/industry standard benchmarks	Amendment 771, Annex VIII, Section C, page 348 and Amendment 399, Article 28b, page 200
	Testing	Report the results of internal and external testing of the foundation model	Amendment 771, Annex VIII, Section C, page 348 and Amendment 399, Article 28b, page 200
Deployment	Machine-generated content	Disclosed content from a generative foundation model is machine-generated and not human-generated	Amendment 101, Recital 60g, Page 76
	Member states	Disclose EU member states where the foundation model is on the market	Amendment 771, Annex VIII, Section C, page 348
	Downstream documentation	Provide sufficient technical compliance for downstream compliance with the EU AI Act	Amendment 1 01, Recital 60g, page 76 and Amendment 399, Article 28b, page 200

- **Uneven reporting of energy use:** Basic model providers report inconsistently on energy use, emissions, their strategies for measuring emissions, and any measures taken to mitigate emissions.

Grading Foundation Model Providers' Compliance with the Draft EU AI Act

Source: Stanford Center for Research on Foundation Models (CRFM), Institute for Human-Centered Artificial Intelligence (HAI)

Draft AI Act Requirements	GPT-4	Cohere Command	Stable Diffusion v2	Claude 1	PaLM 2	BLOOM	LLaMA	Jurassic-2	Luminous	GPT-NeoX	Totals
Data sources	●○○○	●●●○	●●●●	○○○○	●●○○	●●●●	●●●●	○○○○	○○○○	●●●●	22
Data governance	●●○○	●●●○	●●○○	○○○○	●●●○	●●●●	●●○○	○○○○	○○○○	●●●○	19
Copyrighted data	○○○○	○○○○	○○○○	○○○○	○○○○	●●●○	○○○○	○○○○	○○○○	●●●●	7
Compute	○○○○	○○○○	●●●●	○○○○	○○○○	●●●●	○○○○	○○○○	●○○○	●●●●	17
Energy	○○○○	●●○○	●●●○	○○○○	○○○○	●●●●	●●●●	○○○○	○○○○	●●●○	16
Capabilities & limitations	●●●○	●●●○	●●●○	○○○○	●●○○	●●●●	●●○○	●●○○	●●○○	●●●○	27
Risks & mitigations	●●●○	●●○○	●○○○	●○○○	●○○○	●●●○	●●○○	○○○○	○○○○	●●○○	16
Evaluations	●●●●	●●●○	○○○○	○○○○	●●●○	●●●○	○○○○	○○○○	●○○○	●○○○	15
Testing	●●●○	●●●○	○○○○	○○○○	○○○○	●●●○	○○○○	○○○○	●○○○	○○○○	10
Machine-generated content	●●●○	●●●○	○○○○	●●●○	●●○○	●●●○	●●○○	●●●○	●○○○	●●●○	21
Member states	●●○○	○○○○	○○○○	●●○○	●●●●	○○○○	○○○○	○○○○	●○○○	●●○○	9
Downstream documentation	●●●○	●●●●	●●●○	●○○○	●●●●	●●●●	●●○○	○○○○	○○○○	●●●○	24
Totals	25 / 48	23 / 48	22 / 48	7 / 48	27 / 48	36 / 48	21 / 48	8 / 48	5 / 48	29 / 48	

FIGURE 4.1 Evaluation of various model providers for 12 EU requirements on a scale from 0 (worst) to 4 (best). (Human-Centered Artificial Intelligence.)

- **Inadequate disclosure of risk mitigation/non-mitigation**: The risk landscape for core models is immense and encompasses many forms of malicious use, unintended harm, and structural or systemic risk. While many core model vendors list risks, few disclose the mitigations they implement and the effectiveness of these mitigations.

- **Absence of assessment standards/audit ecosystem**: Model vendors rarely measure model performance in terms of intentional damage, such as malicious use, or factors such as robustness and calibration.

In general, no model provider achieves a perfect score, with ample room for improvement in most cases. With sufficient incentives (e.g., fines for non-compliance), companies will change their behavior; even in the absence of strong regulatory pressure, many providers could achieve total scores between 30 and 40 through significant but plausible changes. To be concrete, the maximum entry in OpenAI and Hugging Face is 42 (compliance of almost 90%). Thus, enforcing these 12 requirements in the law would bring a substantial change while remaining within the reach of providers.

In conclusion, model suppliers are unevenly compliant with the requirements set out in the EU AI bill. The enactment and enforcement of the EU AI Bill could bring about significant positive change in the underlying model ecosystem.

4.3 METRICS AND BENCHMARKS

For many benchmarking *datasets* in NLP, metrics exist that allow different LLMs to be compared to each other. Depending on the task, models are evaluated with different metrics. Apart from the traditional metrics (i.e., Accuracy, F1-score, and Perplexity), the most usual metrics used include the following:

- **Exact Matching**: the proportion of predictions that exactly match any of the responses.

- **MMLU** (Multitask Language Understanding): is a reference point (aka benchmark) for large-scale assessment of multitask knowledge understanding. It covers a wide range of knowledge domains from mathematics and computer science to humanities and social sciences.

- **AI2 Reasoning Challenge**: it is a set of school-level science questions.

- **HellaSwag**: it is a common sense inference test that is easy for humans (~95%) but challenging for LLMs.

- **TruthfulQA**: it is a test to measure whether the model can reproduce falsehoods usually found online.

- **TriviaQA**: it is a reading comprehension *dataset* containing over 650,000 triples of questions, answers, and tests. It includes 95,000 question-answer pairs created by trivia enthusiasts and independently collected evidence documents, six per question on average.

- **Bias**: this measures how much bias, and misinformation, exists in the answers generated by different models. This is a subjective assessment produced by human annotators.

- **Model size**: it measures the complexity of an LLM from the point of view of the number of parameters (i.e., millions, billions, trillions...) and/or the internal configuration (i.e., attention layers or hidden layers).

- **GSM8K**: it is a *dataset* of 8.5K high-quality linguistically diverse elementary mathematics verbal problems of high quality created by human problem writers. The solutions mainly involve performing a sequence of elementary computations using basic arithmetic operations to arrive at the final answer.

4.4 BENCHMARK *DATASETS*

Usually, to evaluate NLP tasks with the above metrics, various benchmarks, and *datasets* such as SQuAD, CoCA, and GLUE are used.

4.4.1 SQuAD (Stanford Question-Answering *Dataset*)

This *dataset* was created to advance the area of reading comprehension. Reading text and answering questions about it is a demanding task for machines and requires large, high-quality *datasets*. This one usually contains 107,785 question-answer pairs for 536 Wikipedia articles. For each question, the answer is a text segment, from the corresponding reading passage. The most recent version of SQuAD[2] contains 53,775 new unanswered questions in the same paragraphs.

As a benchmark, humans achieve an EM score of 86.831 and an F1 score of 89.452, leaving these values as a *baseline* for future comparison.

4.4.2 GLUE (General Language Understanding Evaluation)

Most NLP models are designed to solve a specific task, such as answering questions from a particular domain. This limits the use of models for understanding natural language. To process language in a way that is not limited to a specific task, genre, or *dataset*, models must be able to solve a variety of tasks well. GLUE[3] is a set of tools designed to support models that share common linguistic knowledge across tasks. These tasks include textual *entailment*, sentiment analysis, and question answering.

Models tested in GLUE need only have the ability to process single-sentence and sentence-pair inputs and make appropriate predictions. This test suite contains a total of nine sentence or sentence-pair NLU tasks based on established annotated *datasets*. Usually, three different types of tasks are considered in GLUE: single-sentence tasks, similarity and paraphrasing tasks, and inference tasks. For the latter, usually *entailment* (i.e., the hypothesis states something that is definitely correct about the situation in the premise), neutrality (i.e., the hypothesis states something that could be correct about the premise), and contradiction (i.e., the hypothesis states something that is definitely incorrect about the premise) tasks are considered. The collections of sentence pairs that have been created are designed to exemplify various known sources of *entailment* or implicature, from low-level word meanings and sentence structure to high-level reasoning and application of world knowledge (Wang, 2023).

For example, suppose the text, "a car sped past a bunch of people." If the hypothesis sentence were "a man is driving along a quiet road," the class of the inference would be **contradiction**.

Note that the task becomes a classification of sentence pairs by matching each question and each sentence in the respective context, and hence the metric used is Accuracy. Overall, the human reference score is 87.1.

4.4.3 SNLI (Stanford Natural Language Inference)

When it comes to understanding natural language, *entailment* and contradiction inference is essential. The characterization and use of these relations in computational systems is called "natural language inference" or Natural Language Inference (NLI) and is fundamental to tasks such as commonsense reasoning and information retrieval. SNLI[4] is a collection of 570k sentence pairs that are labeled as "*entailment*," "contradiction," or "semantic independence." While there are other data *datasets* that attempt to perform this particular task, they all have problems of size, quality, and vagueness.

The models are again evaluated for the *accuracy* of the predicted label, and there is no human performance measurement for the SNLI corpus.

4.4.4 ARC (Abstraction and Reasoning Corpus)

This repository contains the data of an *Abstraction and Reasoning Corpus* (ARC) task for humans to attempt to solve the tasks manually (Wang, 2023). ARC[5] is a general AI benchmark that is aimed at both humans and computational systems that aim to emulate a form of general human-like general fluid intelligence. An examinee is said to solve a task when, upon first viewing the task, it can produce the correct output for all test inputs in the task (this includes choosing the dimensions of the output). For each test input, the examinee is allowed three attempts (this is valid for all examinees, whether human or AI). The ARC *dataset* includes a collection of 7,787 English science test questions.

4.5 LLM ASSESSMENT

Various LLMs can be evaluated under different criteria and metrics, and using various *datasets*, such as those described previously.

For example, in language inference on SNLI, some models have been evaluated in different inference tasks on different *datasets*, as shown in Table 4.2. The *datasets* used -A1, A2 and A3- correspond to subsets extracted from Wikipedia. This shows that, in general, smaller, fine-tuned BERT models such as RoBERTa outperform even GPT-3.

In language comprehension (MMLU) tasks over multiple domains (i.e., humanities, STEM, social sciences, etc.), some results can be observed in Table 4.3. Note that the values of the metrics depend on the complexity of the model (i.e., number of parameters and training technique), in which case PaLM (540B) is observed to deliver the best performance in all domains. All metrics are exact match accuracy variants, unless otherwise specified.

However, you could already realize that the complexity of each model produces better results, depending on the task being evaluated.

In reading comprehension tasks, the assessment is based on asking about a paragraph or document, and the answer is usually a part of the document. Some specific comprehension tasks include multiple choice, free-form response, paragraph prediction... The result of accuracy assessment in comprehension tasks is shown for some models in Table 4.4, where the best performance is achieved with a smaller BERT model, called ALBERT.

TABLE 4.2   Comparison of LLMs According to Language Inference Capabilities on A1, A2, and A3 Datasets

Model	A1	A2	A3
RoBERTa (Large)	75.5	51.4	49.8
GPT-3	36.8	34.0	40.2
ALBERT	73.6	58.6	53.4
BERT (Large)	57.4	48.3	43.5

TABLE 4.3   MNLI Evaluation for LLMs of Different Complexity

Model	Type	Average (%)	Humanities	STEM	Social Sciences
PaLM 540B	Few-shot	69.3	77	55.6	81
LLaMA 65B	Few-shot	63.4	61.8	51.7	72.9
LLaMA 33B	Few-shot	57.8	55.8	46	66.7
GPT-3	Fine-tuned	53.9	52.5	41.4	63.9
LLaMA 13B	Few-shot	46.9	45	35.8	53.8
BLOOM 176B	Few-shot	39.13	34.05	36.75	41.5
LLaMA 7B	Few-shot	35.1	34	30.5	38.3

TABLE 4.4 Results of the Models in Reading Comprehension Tasks

Model	Accuracy	Accuracy (average)	Accuracy (high)
ALBERT	91.4	–	–
Megatron-BERT	90.9	93.1	90
RoBERTa	83.2	86.5	81.3
PaLM 540B	–	68.1	49.1
PaLM 62B	–	64.3	47.5
PaLM 8B	–	57.9	42.3
LLaMA 65B	–	67.9	51.6
LLaMA 33B	–	64.1	48.3
LLaMA 13B	–	61.6	47.2
LLaMA 7B	–	61.1	46.9
GPT-3 175B	–	58.4	45.5
BLOOM 176B	–	52.3	39.14

TABLE 4.5 Performance of Different LLMs on ARC and QA Tasks

Model	Average (%)	ARC	HellaSwag	MMLU	TruthfulQA
FALCON 40B	63.2	61.6	84.4	54.1	52.5
LLAMA 30B	59.8	58.5	82.9	44.3	53.6
LLAMA 65B	58.3	57.8	84.2	48.8	42.3
LLAMA 13B	51.8	50.8	78.9	37.7	39.9
GPT4 30B	57.9	56.7	81.4	43.6	49.7
GPT4 7B	50.6	48.8	76.6	35.9	41.2
GPT4 13B	53.2	50.8	76.6	38.3	46.9
VICUNA-7B	52.2	47	75.2	37.5	48.9
VICUNA 13B	54.4	50.2	77	40.4	49.8
DOLLY 12B	44.9	41.2	72.3	31.7	34.3
ALPACA 13B	52.7	49.8	79.4	38.9	42.8

In more complex reasoning tasks (ARC), Table 4.5 shows the comparison according to *accuracy* for ARC tasks and QA-related tasks. It can be seen that the FALCON 40B model is the one that obtains the best result for all commonsense reasoning tasks.

On the other hand, many of the results in different tasks can be explained by the complexity and scalability of LLMs, as this is related to their ability to produce emergent skills and is related to the number of parameters that each model must adjust, which can be visualized in Figure 4.2.

Tasks related to QA, inference, textual *entailment*, and others have also been recently evaluated for variants of GPT-4 and BERT (see Table 4.6), where the former outperforms the other models in accuracy.

Some models such as LLaMA and GPT-3 have also been evaluated in relation to biases, misinformation and other aspects found especially in informal texts; for example, Table 4.7 shows the level of bias of LLaMA with respect to GPT-3, where a higher score indicates higher bias, with respect to a CrowSPairs[6] reference *dataset*.

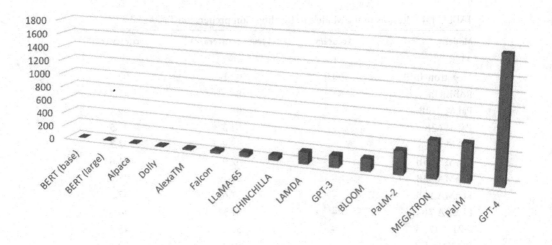

FIGURE 4.2   Comparison of the sizes of LLMs in billions of parameters.

TABLE 4.6   Assessment of Models on GLUE and SQuAD

Model	GLUE	SQuAD
BERT	80.4	93.2
ALBERT	89.4	92.2
GPT-3	93.2	88.8
GPT-4	93.6	89.0

TABLE 4.7   Bias Assessment of LLaMA vs GPT-3

Characteristic	LLaMA	GPT-3
Gender	70.6	62.6
Religion	79.0	73.3
Race	57.0	64.7
Sexual orientation	81.0	76.2
Age	70.1	64.4
Nationality	64.2	61.6
Disability	66.7	76.7
Appearance	77.8	74.6
Socioeconomic status	71.5	73.8
Average	66.6	67.2

Perhaps a recent benchmark that has attracted much attention because of its implications for educational systems is related to the comparison of these models with human performance when specialized standard tests are taken.

Recently, the performance of LLMs such as GPT-3 and GPT-4 was measured based on the performance of various professional and academic exams in the United States[7]. These include the SAT (Scholastic Aptitude Test), the Bar Exam (law), graduate college exams such as the GRE (Graduate Record Examination), and some AP (Advancement Placement) exams (i.e., a college-level testing program for high school students). Performance was measured in percentiles, which were based on the most recent score distributions available

TABLE 4.8   GPT-3 and GPT-4 Test Evaluation Results

Category	Test	GPT-4	GPT-3
Law	Bar Exam	90	10
SAT	Reading and Writing	93	87
SAT	Mathematics	89	70
GRE	Quantitative	80	25
GRE	Verbal	99	63
GRE	Writing	54	54
AP	Calculus	43	0
AP	Physics	66	30
AP	Statistics	85	40
AP	English	14	14

for test takers for each type of exam. This percentile score is a way of ranking a person's performance relative to the performance of others; for example, if you placed in the 60th percentile on a test, this means that you scored higher than 60% of the examinees. Table 4.8 shows some of the results compared to human performance.

As can be seen, GPT-4 is much more capable than GPT-3 in most of these exams. However, it was not able to improve in AP English.

4.6 CONCLUSIONS

In this chapter, we have addressed the crucial importance of having robust assessment measures for LLMs, good benchmarks and *datasets*. These tools are critical for understanding the performance and effectiveness of these models on various language tasks. It has been observed that, although perplexity and accuracy are common metrics, they are not sufficient to fully evaluate the quality of language models.

On the other hand, evaluating LLMs remains an evolving challenge due to their complexity and ability to process large amounts of data. It is critical to develop appropriate *datasets* and design evaluation tasks that accurately reflect the practical use of these models in the real world. In addition, the importance of considering regulatory aspects has been highlighted, to avoid bias and promote fairness in the performance of language models across different user groups.

In general, there are no perfect metrics, as they depend on the tasks and data available; hence he suggests using more than one metric to evaluate models, better human evaluation metrics, more diverse reference data, incorporating diversity metrics and aspects of real evaluation such as robustness.

NOTES

1 https://crfm.stanford.edu/2023/06/15/eu-ai-act.html
2 https://rajpurkar.github.io/SQuAD-explorer/
3 https://super.gluebenchmark.com/
4 https://nlp.stanford.edu/projects/snli/
5 https://github.com/fchollet/ARC
6 https://github.com/nyu-mll/crows-pairs
7 https://www.visualcapitalist.com/how-smart-is-chatgpt/

## CHAPTER 5

Applications

5.1 INTRODUCTION

LLMs have revolutionized the field of NLP by showing extraordinary power to understand and generate text in a variety of tasks. These models have opened the door to a wide spectrum of applications ranging from virtual assistants to machine translation and sentiment analysis. In this chapter, we will explore several practical examples for various tasks, such as sentiment classification, semantic search, *fine-tuning* for question-answer tasks, interfaces for database access, causal inference, autonomous models, and summary generation, among others, using different available LLMs.

All the examples discussed are developed in Python on Google's Colab platform. The source code and data used in these examples can be downloaded directly from the publisher's website. Several exercises assume some familiarity with basic aspects of machine learning and frameworks such as TensorFlow and Keras. In addition, in several exercises, you will be required to have an API access key from some vendors such as OpenAI. Note that, depending on the characteristics of the LLM, some examples are based on Spanish texts and others on English.

5.2 SENTIMENT CLASSIFICATION

For this example, you should use the program *BERT_Sentiment_Classifier* and the *dataset* `"sentiments-english.xls"`. The program allows you to create an automatic sentiment classifier from opinions collected from social networks. These opinions are contained in a two-column OLS file: the opinion and the polarity of the sentiment (i.e., positive, negative).

To achieve the above, two steps are performed:

1. *Fine-tuning* the pre-trained BERT model with our own data (spreadsheet with opinions and their polarities), so to generate the final contextual *embeddings*. For this purpose, the bert_en_uncased_L-24_H-1024_A-16 model is used, which has L = 24 *transformer* blocks, H = 1024 hidden layers and A = 16 *attention heads*.

DOI: 10.1201/9781003517245-5

2. Building the complete model for which, to the output of the fitted BERT model, we add layers of a simple neural network that allows to predict the polarity (positive, negative) of the output (i.e., *embedding*) of the pre-trained BERT model.

First, we install some packages that allow working with the TensorFlow framework, *transformers*, *tokenization*, and CSV file formatting:

```
!pip install tensorflow
!pip install bert-for-tf2
!pip install tokenization
!pip install transformers
!pip install --upgrade xlrd
```

Then, we import libraries to use *transformers*, *tokenizers*, neural network layer handling (Sequential, Model, Dense, Input) and other general purpose ones:

```
from transformers import InputExample, InputFeatures
from bert.tokenization.bert_tokenization import FullTokenizer
from bert import tokenization
from keras.models import Sequential
from keras.models import Model
from keras.layers import Dense
from keras.layers import Input
import tensorflow     as tf
import tensorflow_hub as hub
from tensorflow.keras.optimizers import Adam
import numpy as np
import pandas as pd
import torch
from sklearn.model_selection import train_test_split
```

We define a function to load the BERT model from one of several repositories: TensorHub:

```
Def LoadBERT():
 module_url = "https://tfhub.dev/tensorflow/bert_en_uncased_L-24_H-
1024_A-16/1"
 bert_layer = hub.KerasLayer(module_url, trainable=False)
 return(bert_layer)
```

We define a function to load the *tokenizer*, where:

- vocab_file: is a vocabulary file to transform our own *dataset* into *feature* vectors (*embeddings*).

- do_lower_case: method that converts the *tokens* generated by the *tokenizer* to lowercase:

```
def LoadTokenizer():
    vocab_file =
        bert_layer.resolved_object.vocab_file.asset_path.numpy()
    do_lower_case = bert_layer.resolved_object.do_lower_case.numpy()
    tokenizer = FullTokenizer(vocab_file, do_lower_case)
    return(tokenizer)
```

Recall that BERT requires formatting the input text sequence at three levels: *embedding* of each *token*, positional *embedding* and *embedding* of each segment, as shown in the figure, with the corresponding sentence separator labels: [CLS] (sentence start) and [SEP] (separator).

We define the function **bert_encoder(texts, *tokenizer*, max_len)** that takes a list of texts (i.e., opinions), the ***tokenizer*** and the maximum length of the sentences (**max_len**) and returns the encoding required for training: *input tokens, masks, and segments.*

For this, the following steps must be performed:

- To *tokenize* each sentence by converting it to a sequence of identifiers for each word.

- To add special *tokens* to separate sentences and perform the classification: sentence onset (CLS) and separators (SEP).

- To convert to fixed-length sequences (512), so **padding** (PAD) must be performed. This is because BERT works with fixed-length sequences (i.e., 512), so the difference for shorter sequences must be "padded" with 0s.

```
def bert_encoder(texts, tokenizer, max_len=512):
    all_tokens = []
    all_masks = []
    all_segments = []
    for text in texts:
        text = tokenizer.tokenize(text)
        # Cut two words in case sentences have more than
        # 512 characters long
        text_sequence = text[:max_len-2]
        # Add sentence start (CLS) and separation (SEP) tags
        input_sequences = ["[CLS]"] +text_sequence+ ["[SEP]"]
        # Calculate the length for padding
        pad_len = max_len - len(input_sequences)
        # Convert tokens into BERT internal IDs
        tokens =
            tokenizer.convert_tokens_to_ids(input_sequences)
        # Add pad_len tokens for  padding (0s) and masks for
        # input sequences and segments
        tokens += [0] *  pad_len
        pad_masks = [1]*len(input_sequences)+[0]*pad_len
```

```
        segment_ids = [0] * max_len
        # Add tokens, padding masks and segments to lists
        all_tokens.append(tokens)
        all_masks.append(pad_masks)
        all_segments.append(segment_ids)
    return np.array(all_tokens), np.array(all_masks), np.array(all_
segments)
```

Next, we define the function **BuildModel (bert_layer, num_classes, max_len),** which builds a classification model for input sentences of length and classes (**max_len** and **num_ classes**) using BERT's previous layers (**bert_layer**). For this, we use the pre-trained BERT model and add some prediction layers to the neural network, which will perform sentiment classification.

Specifically, we create a network with a *dense layer*, with an output that predicts the probability that the opinions are either 1 (positive) or 0 (negative), using a binary classifier. The error (loss) is simply measured with a *binary_cross_entropy* metric.

Each *layer* receives an array of three elements, of which two *embeddings* are used for training [[**input_words_*tokens*]][**input_masks**][**segment_ids**]].

Finally, we need to create three input layers (words, masks and segments) of size **max_len**:

```
def BuildModel(bert_layer, num_clases,max_len=512):
    input_word_ids      = Input(shape=(max_len,), dtype=tf.int32,
name="input_word_ids")
    input_mask          = Input(shape=(max_len,), dtype=tf.int32,
name="input_mask")
    segment_ids         = Input(shape=(max_len,), dtype=tf.int32,
name="segment_ids")
    # Merge input layers (word IDs, masks, segments)
    _, output_sequence = bert_layer([input_word_ids, input_mask,
segment_ids])
    classifier_output  = output_sequence[:, 0, :]
    # Dense layer classifies a polarity from the BERT output layer
    output_layer  = Dense(num_clases,
activation='sigmoid')(classifier_output)
    model         = Model(inputs=[input_word_ids, input_mask, segment_
ids], outputs=output_layer)
    model.compile(Adam(learning_rate=2e-6), loss='binary_
crossentropy', metrics=['accuracy'])
    return model
```

Now, we start the main program, loading the BERT model and the *tokenizer*

```
bert_layer = LoadBERT()
tokenizer  = LoadTokenizador()
```

We load a file (*"sentiments-ingles.csv"*) in English, with examples of opinions and their sentiment polarities, replacing the latter by numeric values (*Negative = 0, Positive = 1*):

```
opinions=pd.read_csv('/content/sentiments-english.csv',
                     header=None)
opinions[1].replace(['Negative','Positive'],
                    [0,1],inplace=True)
```

We separate the opinion *dataset* into a subset of training (*train*) and test (*test*):

```
train, test = train_test_split(opinions)
```

Then, we encode the training data with a maximum length of 160 characters per sentence (160 is calculated as the maximum length possessed by the sentences in the example *dataset*):

```
train_input = bert_encoder(train[0], tokenizer, max_len=160)
```

We build the model with the **bert_layer** for a class (0 or 1):

```
Final_model = BuildModel(bert_layer, 1,160)
```

5.2.1 Training

Now we perform the *fine-tuning* of the model with the input pre-coding and with the class labels (*train*[1]). For this, a separation of the training data (*split*) is performed, where 20 % will correspond to the data used for validation (*validation_split*). In addition, it is trained with a certain number of epochs and batch set size:

```
train_history = Final_model.fit(
    train_input, train[1],
    validation_split=0.2,
    epochs=3,
    batch_size=16
    )
```

5.2.2 Testing and Validation

Remember that, for model testing, the test data **(test[0])** must have the same format as the training data. For this, we invoke our **bert_encoder(..)** function again with the test data to convert it into three *embeddings*, which will be passed to the classification method (i.e., positional, segment and *embedding*).

ProbPrediction is an array containing the classification probability of each opinion. If the probability of classification is greater than a threshold (50%), the class is positive (1); otherwise, it is negative (0):

```
test_input = bert_encoder(test[0], tokenizer, max_len=160)
ProbPrediction  = Final_model.predict(test_input)
```

```
prediction = np.where(ProbPrediction>.5, 1,0)
prediction
```

5.3 SEMANTIC SEARCH

For this exercise, you should use the program SemanticSearch, which allows you to perform semantic search using the *embedding* of a user *query* and delivering the most similar texts of a *dataset*, based on cosine similarity measures. The *embeddings* are obtained from multiple GPT3-5 *embedding* models, available from OpenAI (i.e., text-*embedding*-ada-002).

First, we install some packages to access the models:

```
!pip install openai
```

We import some libraries and methods from OpenAI to obtain *embeddings* and compute cosine similarity between vectors:

```
import pandas as pd
import numpy as np
import pprint
import openai
from openai.embeddings_utils import get_embedding, cosine_similarity
```

We define the function **SearchComments(Opinions,Query,n)**, which searches for similarities (cosine) between a *query* and **Opinions** and delivers the **n** best results. For this, the *query* is converted to its corresponding *embedding*:

```
def SearchComments(Opinions, Query, n=3):
    QueryEmbedding = get_embedding(
        Query,
        engine=EMBEDDING_MODEL
    )
    # Calculate cosine similarity between query's embedding and
    # all opinions' embeddings
    Opinions["sim"] = Opinions.embedding.apply(
            lambda x: cosine_similarity(x, QueryEmbedding))
    # Sort similarity results
    results = Opinions.sort_values("sim", ascending=False).head(n)
    return results
```

We start the main program:

```
openai.api_key = "INSERT API KEY"
# Select an embedding model
EMBEDDING_MODEL = "text-embedding-ada-002"
Opinions = pd.read_csv('/content/sentiments-english.csv', header=
None)
Opinions["embedding"] = Opinions[0].apply(lambda x: get_embedding
(x, engine=EMBEDDING_MODEL))
```

We perform the search for a given *query*:

```
results = SearchComments(Opinions, "borderlands quite", 5)
print(results[0])
589          Borderlands stuff is quite fun
587        Borderlands 3 is quite different
584          Borderlands 3 is quite fun
588     Borderlands 3 is actually quite fun
585          Borderlands 3 is really fun
```

5.4 REASONING WITH LANGUAGE AGENTS

In this example, you should use the `LangchainAgents` program, which allows you to use OpenAI's LangChain *framework* to develop simple LLM-based applications. LangChain allows to connect an LLM to other data sources (i.e., SQL databases or Google search engine) and allows the LLM to interact with its environment using Agents.

LangChain allows generating step-by-step reasoning chains to respond to high-level tasks, decomposing them into simpler tasks. To achieve this, an agent has access to several tools and determines which one to use, depending on the user's input. In general, there are two types of agents:

1. **Action agents**: They decide the next action, using the outputs of previous actions. These are suitable for small tasks.

2. **Plan-and-execute agents**: They decide on the complete sequence of actions and then executes them all. These are suitable for tasks that require maintaining long-term goals.

We install some packages such as LangChain and Google search engines:

```
!pip install  langchain openai pymysql --upgrade -q
!pip install  google-search-results -q
```

We import some LangChain libraries for language agent use and set the environment variables for use of the respective OpenAI (OPENAI_API_KEY) and Google (SERPAPI_API_KEY) APIs, for which you must get the respective *keys* to use, such as the Google search engine (serpAPI) and an LLM (by default, a pre-trained GPT model such as "text-davinci-003"):

```
from langchain.agents import load_tools
from langchain.agents import initialize_agent
from langchain.agents import AgentType
from langchain.llms import OpenAI
import os
os.environ['OPENAI_API_KEY']  = "INSERT OpenAI API KEY"
os.environ["SERPAPI_API_KEY"] = " INSERT Google Search API KEY"
```

We initialize the OpenAI libraries for LLM and some tools:

```
llm = OpenAI()
tools = load_tools(["serpapi", "llm-math"], llm=llm)
```

In this example, we use a simple language agent without memory (*ZERO_SHOT_REACT_DESCRIPTION*); that is, the action it performs is based only on the current action and not on previous ones (history). Thus, the agent decides which tool to use based solely on the tool description:

```
agent = initialize_agent(tools, llm,
agent=AgentType.ZERO_SHOT_REACT_DESCRIPTION)
agent.run("Who is the president of croatia's wife and how old will be
she  in 10 years?")
```

```
The president of Croatia's wife will be 64 years old in 10 years.
```

Now, we ask the agent to show step-by-step what it performed (verbose). In this case, the plan-and-execute type agent performs the following:

1. It receives an input from the user.

2. It plans the complete sequence of steps to take.

3. It executes the steps in order, passing the outputs of the previous steps as input to the future steps.

Usually, the planner is an LLM, and the executor is an action:

```
agent = initialize_agent(tools, llm,
agent = AgentType.ZERO_SHOT_REACT_DESCRIPTION)
agent.run("Who is the president of croatia's wife and how old will be
she  in 10 years?")
```

```
The president of Croatia's wife will be 64 years old in 10 years.
```

Then, the chain of reasoning for the answer begins:

```
> Entering new  chain...
I need to first find out who the spouse of the president of Croatia
is and then calculate what age she will be in 10 years.
Action: Search
Action Input: "President of Croatia spouse"
Observation: Sanja Musić Milanović
Thought: Now I need to calculate her age in 10 years
Action: Calculator
```

```
Action Input: Sanja Music Milanovic's age +10
Observation: Answer: 40
Thought: I now know the final answer
Final Answer: Sanja Music Milanovic will be 40 years old in 10 years.
```

```
> Finished chain.
```

5.5 CAUSAL INFERENCE

For this exercise, you should use the program `CausalLM _ LLaMA`, which performs causal inference on the open source LLaMA LLM of 3B parameters, and pre-trained on 1T *tokens*.

Causal inference on LLMs allows understanding cause-effect relationships between variables (aka CausalLM). This can be applied to answer (effect) a question (cause), to establish *entailment* between one sentence (cause) and another (effect), etc. In this example, the causal model will be used to answer questions. The LLaMA model can use either its own *token*izer (LLama*Token*izer) or the generic Auto*Token*izer method.

First, we installed some *transformer* packages and facilities for hardware acceleration:

```
!pip install transformers
!pip install sentencepiece
!pip install accelerate -U
```

We import some libraries for *transformers*:

```
from transformers import LlamaTokenizer, LlamaForCausalLM,
AutoTokenizer
from transformers import TrainingArguments,
AutoModelForSequenceClassification
import torch
import numpy as np
```

We define the function **LoadModel(ModelName)** to load the *token*izer (LLama*Token*izer) and the pre-trained model, based on **ModelName**, returns the loaded model and *token*izer:

```
def LoadModel(ModelName):
    tokenizer = LlamaTokenizer.from_pretrained(ModelName)
    model = LlamaForCausalLM.from_pretrained(
        ModelName, torch_dtype=torch.float16, device_map='auto')
    return(model,tokenizer)
```

We define the function ***FormatPrompt(prompt,tokenizer)***, which *token*izes a string indicating the *prompt*** and returns the IDs of each *token* that composes it:

```
def FormatPrompt(prompt,tokenizer):
    inputIDs = tokenizer(prompt, return_tensors="pt").input_ids
    return(inputIDs)
```

We start the main program, loading the model, defining the *prompt* to send to the model and generating the output with a maximum number of *tokens*:

```
# Other LLaMA models:
#      'openlm-research/open_llama_7b'
#      'openlm-research/open_llama_13b'
(model,tokenizer) = LoadModel('openlm-research/open_llama_3b')
# The prompt must start with  Q (question) and end with
A (Answer)
# so that the model infers what it follows to  A
prompt   = 'Q: What is the biggest animal in Chile?\nA:'
inputIDs = FormatPrompt(prompt,tokenizer)
output   = model.generate(input_ids=inputIDs, max_new_tokens=20)
```

Note that the model generates an output response with the IDs of each *token*, so these must be decoded (*decode*) to convert them into words:

```
print(tokenizer.decode(output[0]))

<s>Q: What is the biggest animal in Chile?
A: The wolf
```

5.6 NATURAL LANGUAGE ACCESS TO DATABASES

In this application, you must use the SQL _ Query, program, which allows you to interpret natural language *queries* in SQL databases through models such as GPT. For this, the following steps are followed:

1. Load a table with sales data ("sales.csv").

2. Convert the table to SQL.

3. Create a *prompt* to perform a GPT *query* in the SQL database.

4. Return the answer.

First, we install some packages to access the models:

```
!pip install openai
```

We import some general purpose libraries and some libraries to handle SQL databases (*sqlalchemy*):

```
import pandas as pd
import numpy as np
import pprint
import openai
```

```
from sqlalchemy import create_engine
from sqlalchemy import text
```

We define a function to initialize the SQL database with the data from the sales table "*sales.csv*":

```
def InitDB():
    temp = create_engine("sqlite:///:memory:", echo=True)
    data = df.to_sql(name="Sales", con=temp)
    return(data)
```

We define a **Create***Prompt***(df)** function that tells GPT about the **df** data and its properties, which we will use (a table is created with all the columns of the initial data):

```
def CreatePrompt(df):
  prompt = '''### sqlite SQL table:
#
# Sales({})
#
'''.format(",".join(str(x) for x in df.columns))
    return(prompt)
```

We create a function Combine*Prompt*(df,query) to combine the user *query query* **consulta** with the df table structure with the additional string "**A query to reply**:", followed by the keyword "**Select**," so that GPT understands the *query* correctly:

```
def CombinePrompt(df,query):
    defi = CreatePrompt(df)
    query_string = f'### A query to reply:
                    {query}\nSELECT'
    return(defi+query_string)
```

We defined the function **GenerateGPTResponse(df,nlp_text)** to invoke the OpenAI API and use the "*text-davinci-003*" model to deliver the results, as well as some other parameters, such as the temperature and maximum number of *token*s to return:

```
def GenerateGPTResponse(df,nlp_text):
    response = openai.Completion.create(
      engine ="text-davinci-003",
      prompt = CombinatePrompt(df,nlp_text),
      max_tokens = 150,
      n=1,
      stop=['#',';'],
      temperature=0.7,
    )
    return(response)
```

We create the function **HandleResponse(answer)** that parses the response and passes it to the database:

```
def HandleResponse(answer):
    query = answer.choices[0].text
    if query.startswith(" "):
        query = "Select"+query
    return(query)
```

Now, we start the main program:

```
openai.api_key = "INSERT API KEY"
df = pd.read_csv('/content/sales.csv', encoding="latin1")
data = InitDB()
nlp_text = input("Input query:")
response = GenerateGPTResponse(df,nlp_text)
Input query: I want to know which products sold the most?
print(response.choices[0].text)
```

We show the processing that the model performs:

```
PRODUCTLINE, SUM(QUANTITYORDERED) AS TOTAL_QUANTITY
FROM Sales
GROUP BY PRODUCTLINE
ORDER BY TOTAL_QUANTITY DESC
```

We connect to the SQL server to respond to the SQL *query*:

```
with temp.connect() as conn:
    result = conn.execute(text(HandleResponse(response)))
```

The SQL manager response:

```
INFO sqlalchemy.engine.Engine Select PRODUCTLINE,
SUM(QUANTITYORDERED) AS TOTAL_QUANTITY
FROM Sales
GROUP BY PRODUCTLINE
ORDER BY TOTAL_QUANTITY DESC

INFO:sqlalchemy.engine.Engine:Select PRODUCTLINE,
SUM(QUANTITYORDERED) AS TOTAL_QUANTITY
FROM Sales
GROUP BY PRODUCTLINE
ORDER BY TOTAL_QUANTITY DESC
```

Finally, we show the results of the SQL *query*:

```
result.all()
```

```
[('Classic Cars', 33992),
 ('Vintage Cars', 21069),
 ('Motorcycles', 11663),
 ('Trucks and Buses', 10777),
 ('Planes', 10727),
 ('Ships', 8127),
 ('Trains', 2712)]
```

5.7 LOADING AND *QUERYING* FOR OWN DATA

For this exercise, you should use the program `QueryingOwnData`. This allows *querying* own textual data to an LLM based on comparison of *embeddings* (question-answer or QA). For this purpose, the data to be queried are transformed into vector databases that efficiently store the text *embeddings*. These databases are designed to perform similarity searches in high dimensional spaces, enabling the retrieval of the most semantically relevant results.

By storing document (or paragraph) *embeddings* in a vector database, a search system can quickly identify the texts that best match a *query*.

For this example, OpenAI's LangChain framework will be used to develop simple LLM-based applications. LangChain allows to connect an LLM to other data sources (i.e., SQL databases or Google search engine).

First, we install the necessary LangChain, OpenAI and other packages:

```
!pip install langchain
!pip install duckdb
!pip install unstructured
!pip install chromadb
!pip install BeautifulSoup4
!pip install openai
!pip install tiktoken
```

We import some relevant libraries to get *embeddings* and *query* or retrieve (Retrieve) from the *embeddings* databases:

```
from langchain.document_loaders.unstructured import
UnstructuredFileLoader
from langchain.text_splitter import CharacterTextSplitter
from langchain.embeddings import OpenAIEmbeddings
from langchain.vectorstores import Chroma
from langchain.chains import RetrievalQA
from langchain.chat_models import ChatOpenAI
import openai
import os
import tiktoken
```

We assign the key to be able to access the API of an LLM:

```
os.environ['OPENAI_API_KEY']="INSERT API KEY"
```

We load a test text ("*sample.txt*"):

```
loader = UnstructuredFileLoader('sample.txt')
documents = loader.load()
```

Now, we separate the text into chunks (*chunks*) of 1000 characters (without overlapping chunks: *chunk_overlap*=0):

```
text_splitter = CharacterTextSplitter(chunk_size=1000, chunk_
overlap=0)
MyTexts = text_splitter.split_documents(documents)
```

We initialize the function that will calculate the *embeddings*:

```
MyEmbeddings = OpenAIEmbeddings()
```

We create a vector database from our data and use it to index the *embeddings*:

```
db = Chroma.from_documents(MyTexts, MyEmbeddings)
```

Our data are already indexed (i.e., paragraph with their corresponding IDs and *embeddings*) so that we can *query* them. For this, we use LangChain's **RetrievalQA** function, which initializes our LangChain framework with the LLM available by default ("*text-embedding-ada-002*") and our vector database (db). Note that only one output response is allowed (k = 1):

```
qa = RetrievalQA.from_chain_type(
          llm=ChatOpenAI(),
          chain_type="stuff",
          retriever=db.as_retriever(
          search_kwargs={"k": 1}))
```

Now, we can perform *queries* to our model. For this, we internally convert the *query* to its *embedding*, and look for similarities between it and those available in the vector database of our own data:

```
query = "What is the document about?"
qa.run(query)
```

```
The document is about the resignation of General Director of
Carabineros by Bruno Villalobos Krumm and mentions the plans to
```

```
modernize and strengthen the police forces in the fight against
crime, drug trafficking and terrorism.

query = "Who is Bruno Villalobos?"
qa.run(query)

Bruno Villalobos is the General Director of Carabineros, Chile's
police force. However, in the given context, it is mentioned that
he has resigned from his position.
```

5.8 *FINE-TUNING* A MODEL WITH OWN DATA

For this exercise, you should use the program FALCON _ FineTuning, which performs *fine-tuning* of the pre-trained FALCON model with our own data based on instructions to the model.

The types of pre-trained Falcon models include:

- *Normal:* 'falcon-7b' and 'falcon-40b,' of 7B and 40B parameters, respectively.

- *Instruction-based*: 'falcon-7b-instruct' and 'falcon-40b-instruct'''*, of 7B and 40B parameters, respectively.

- *Optimized for Quantization*: these models are enhanced to run more efficiently on high-performance hardware (aka 4-bit or 8-bit hardware). Quantization is the process of transforming infinite continuous values into a finite set of discrete values, which saves memory. Available models are: 'falcon-7b-instruct-4bit' and 'falcon-40b-instruct-4bit,' with 7B and 40B parameters respectively.

The *fine-tuning* is performed with a JSON format *dataset*, where each training input contains three columns: Instruction (i.e., *prompt* given to the model), Input (i.e., input context information -optional-) and Output (i.e., expected output), as in the following example:

```
{
"instruction": "Classify the following into animals, plants and
minerals",
"input": "Oak tree, copper mine, elephant",
"output": "Oak: Plant Copper mine: Mineral: Elephant: Animal"
}
```

Note that specialized computational resources (i.e., GPU or TPU) are used, so the program in the training or *fine-tuning* stages may take some time. However, you must still have adequate computational resources for this example to work.

First, let's load the *fine-tuning* program (*falcontune.git*) from an external site, and clone it into our store folder:

```
!git clone https://github.com/rmihaylov/falcontune.git
```

Now we load the *embeddings* of the 40B FALCON pre-trained model (with Quantization) into our store folder:

```
!wget https://huggingface.co/TheBloke/falcon-40b-instruct-GPTQ/
resolve/main/gptq_model-4bit--1g.safetensors
```

Then, we move to the program folder (*falcontune*), to install both the model and specific programs for massive computational resource usage (i.e., CUDA, Triton):

```
!cd falcontune && pip install -r requirements.txt
!cd falcontune && python setup.py install
```

We define the name of the models, *embedding* and the *dataset* to perform the *fine-tuning*. Note that the *dataset* in JSON format must be loaded in your store folder ("*MisDatos.json*"):

```
import os

os.environ['BASE_MODEL'] = 'falcon-40b-instruct-4bit'
os.environ['EMBEDDINGS'] = "./gptq_model-4bit--1g.safetensors"
os.environ['DATASET']   = "./MyData.json"
os.environ['TUNED_MODEL'] = "falcon-40b-instruct-4bit-alpaca/"
```

First, we test our model with the *falcontune* program to generate a response for an input *prompt*:

```
!falcontune generate \
    --model    ${'BASE_MODEL'} \
    --weights ${'EMBEDDINGS'} \
    --max_new_tokens 2000 \
    --use_cache \
    --do_sample \
    --prompt "How to prepare a strategic plan?" \
    --backend triton
```

```
How to prepare a strategic plan?

To prepare a strategic plan, you should consider your company's
objectives, external environment, threats, opportunities,
strengths, and weaknesses. Then, you can use various techniques,
such as SWOT analysis, to identify the factors that may affect the
company's performance. Next, you should determine the strategies
to achieve the objectives, assign specific tasks to each area of
the company and establish an implementation schedule. Finally, you
should review and update the strategic plan on a regular basis to
adapt it to market changes.
```

We prepare to perform *fine-tuning* with our previous *dataset*. For this, for now we disable **WanDB**, which is a development platform that allows users to monitor and visualize various aspects of the training in real time:

```
os.environ["WANDB_DISABLED"] = "true"
```

We started to perform the *fine-tuning* (*finetune*) with our *falcontune* program, but now we pass to it the model usable for tuning, the *embeddings,* and the *dataset* itself, as well as other parameters specific to the tuning process. To perform this process more efficiently, LoRA is used, which is a training method that accelerates the training of large models while consuming less memory. The output of the training fitting process remains in *TUNED_MODEL*:

```
!falcontune finetune \
    --model ${'BASE_MODEL'} \
    --weights  ${'EMBEDDINGS'} \
    --dataset  ${'DATASET'} \
    --data_type=alpaca \
    --lora_out_dir  ${'TUNED_MODEL'} \
    --mbatch_size=1 \
    --batch_size=3 \
    --epochs=2 \
    --lr=3e-4 \
    --cutoff_len=256 \
    --lora_r=8 \
    --lora_alpha=16 \
    --lora_dropout=0.05 \
    --warmup_steps=5 \
    --save_steps=30 \
    --save_total_limit=3 \
    --logging_steps=5 \
    --target_modules='["query_key_value"]' \
    --backend=triton
```

Finally, we can test our fitted model with some instruction (i.e., "How to prepare a strategic plan?"):

```
!falcontune generate \
    --model ${'BASE_MODEL'} \
    --weights ${'EMBEDDINGS'} \
    --lora_apply_dir ${'TUNED_MODEL'} \
    --max_new_tokens 2000 \
    --use_cache \
    --do_sample \
    --instruction "How to prepare a strategic plan?"\
    --backend triton
```

1. Define the objectives: It is important to clearly define the objectives of the strategic plan. They should be specific, measurable, achievable, relevant, and timely.
2. Conduct an external and internal assessment: An external assessment should be conducted to determine the environmental, economic, political, social, and technological factors that may affect the organization. It is also necessary to conduct an internal assessment to determine the organization's strengths, weaknesses, threats, and opportunities.
3. Develop strategies: From the above assessments, specific and measurable strategies should be developed to achieve the objectives. These should include objectives and goals, actions to be taken, responsibilities, resources needed and time frames.
4. Implement the plan: Once the plan has been developed, it should be implemented. This includes assigning responsibilities to specific teams and departments, setting deadlines and monitoring progress.
5. Review and adjust: It is important to review and adjust the plan regularly. This includes conducting evaluations, making changes if necessary, and ensuring that the objectives are achievable.

In summary, the steps in developing a strategic plan are define the objectives, conduct an external and internal assessment, develop strategies, implement the plan, and review and adjust. It should be noted that strategic plans vary by organization and industry, so it is important to adapt them to specific needs.

5.9 PROMPT DESIGN AND OPTIMIZATION

There are several applications of conversational systems based on LLMs, such as ChatGPT and BARD. The greatest benefit that can be obtained from these applications depends on the way in which the instructions or *prompts* for the task we want you to perform are provided. LLMs, such as GPT-4 or Palm-2, have been trained in a variety of ways to allow interaction with humans. One of these ways is based on training by instructions, which consists of the following components:

- *Instruction*: a specific task or instruction that you want the model to perform.

- *Context*: may involve external information or additional context, which can direct the model to better responses.

- *Input data*: the input or question for which we are interested in finding an answer.

- *Output Prompt*: indicates the type or format of the output.

Note that not all components are required for a request and the format depends on the task at hand.

For example, a training trio might include:

- **Instruction**: "Classify the following into animals, plants, and minerals."
- **Input**: "roses, copper, cougar."
- **Output**: "Roses: plant, copper: mineral, puma: animal."

Naturally, we could ask questions to applications like ChatGPT, which have no context, and only *prompts*, such as the following:

 You

When were the balkan wars held?

 ChatGPT

The term "Balkan Wars" typically refers to two separate conflicts that occurred in the Balkan Peninsula during the early 20th century. These conflicts were:

1. First Balkan War: October 1912 to May 1913
2. Second Balkan War: June 1913 to August 1913

The First Balkan War was fought primarily between the Balkan League (comprising Serbia, Montenegro, Greece, and Bulgaria) and the Ottoman Empire. The Balkan League sought to gain territory from the Ottoman Empire, and the conflict resulted in significant territorial changes in the region. ↓

Or some factual questions:

 You

What is the capital and population of Croatia?

 ChatGPT

As of my last knowledge update in January 2022, the capital of Croatia is Zagreb. Please note that there might be changes after this date. As for the population, the estimated population of Croatia is around 4 million people. For the most up-to-date and accurate information, I recommend checking the latest data from official sources or recent demographic statistics.

However, we could design more sophisticated *prompts* for ChatGPT or Bard, which consider more contextual information, for which there are some interesting approaches that improve our communication with these types of applications:

Specifying output format: one can specify the format of a response as if one were giving an instruction to a student; for example:

 You
Make a table containing the three most countries where spanish is spoken

 ChatGPT
Here's a table listing the three most populous countries where Spanish is spoken:

Rank	Country	Population (Approx.)
1	Mexico	126 million
2	Colombia	51 million
3	Argentina	45 million

Please note that population figures are approximate and may have changed since my last knowledge update in January 2022. It's always a good idea to refer to the latest demographic data for the most accurate information.

Providing explicit constraints to an answer: we can deliver *prompts* with constraints for more open-ended questions; for example:

 You
Make a summary with of the main event held in the second world war sorted chronologically, in no more than 4 lines

 ChatGPT
1. **1939**: Germany invades Poland, prompting the start of World War II.
2. **1941**: Japan attacks Pearl Harbor; Germany invades the Soviet Union.
3. **1944**: Allies launch D-Day, invading Normandy; Soviets advance into Eastern Europe.
4. **1945**: Allies defeat Germany; U.S. drops atomic bombs on Hiroshima and Nagasaki, leading to Japan's surrender, marking the end of World War II.

Explaining answers with a certain level of difficulty: we can request that the answer possess certain styles and/or level of knowledge; for example:

You
Explain a language model for a 18-year boy in one paragraph

ChatGPT
A language model is like a super-smart virtual assistant for your computer or phone. Imagine having a friend who knows a bit about everything and can help you with homework, answer questions, or even come up with creative writing ideas. This friend, the language model, has read and learned from tons of books, articles, and websites, so it knows a wide range of topics. You can ask it anything, and it will try its best to give you a helpful and informative response. It's like having your own personal guide in the world of words and information, making your online experiences more interesting and engaging.

Providing context to the prompt: we can provide context input information to perform the task; for example:

You
Translate this text into Spanish "A language model is like a super-smart virtual assistant for your computer or phone. Imagine having a friend who knows a bit about everything and can help you with homework, answer questions, or even come up with creative writing ideas."

ChatGPT
Un modelo de lenguaje es como un asistente virtual súper inteligente para tu computadora o teléfono. Imagina tener un amigo que sabe un poco de todo y puede ayudarte con la tarea, responder preguntas o incluso sugerir ideas creativas para escribir.

Acting in a certain role: we can get the system to act in a certain role for our subsequent interactions; for example:

You

Act as an empathic person to serve a customer from now on

ChatGPT

Of course, I'm here to help and provide support. Please feel free to share your thoughts, concerns, or any questions you may have, and I'll do my best to assist you. Your satisfaction is my priority, and I'm here to ensure your experience is positive and fulfilling. How can I assist you today?

🗍 ⬆ ⬇

You

I didn't receive my package

ChatGPT

I'm sorry to hear that you haven't received your package. I understand how frustrating that can be. Let's work together to resolve this issue.

Firstly, have you checked the tracking information for your package? If you have a tracking number, you might be able to see where the package is currently located and get an estimate of when it will be delivered. If you haven't checked it yet, please do so, and let me know if there are any updates.

If you've already checked the tracking information and it doesn't provide clarity or if there's an issue, I recommend reaching out to the shipping carrier or the company from which you made the purchase. They may be able to provide additional details or initiate an investigation into the whereabouts of your package.

If you have any relevant details like the tracking number or the name of the shipping

Prompting with no prior training (zero-shot): we can give instructions or questions with zero prior training; for example:

You

Classify this text into neutral, positive or negativa "there is no better place than Croatia"

ChatGPT

The given text "there is no better place than Croatia" can be classified as positive. It expresses a favorable opinion about Croatia, suggesting that the speaker believes it to be the best place.

🗍 ⬆ ⬇ ⟳

Prompt with few shots: can be used as a technique to enable learning in context, where we provide demonstrations at the *prompt* to guide the model to better performance.

You

A "complete" is a sandwich cooked in Chile and is mainly made of mayonnaise, tomate and avocato. An example of a sentence using completo is: "I was starving yesterday so I ate a completo in Santiago downtown". Give an example of a sentence using completo:

ChatGPT

"During my visit to Chile, I couldn't resist trying the local cuisine, so I ordered a completo at a street vendor, savoring the delicious combination of mayonnaise, tomato, and avocado."

You may have noticed with these examples that the results may vary depending on the length and, therefore, the context provided to the *prompts*. In general, better results can be obtained by showing the reasoning that leads to a specific answer. This is called a "chain of thoughts" (Chain of Thoughts or CoT) and can induce a "think step-by-step" type of behavior. CoT can take the form of questions and intermediate answers (Wei et al., 2023); for example, "Imagine you are a physics professor. Answer this question:". If you rerun the same *query* multiple times, you will likely get different answers, so choosing a consistent answer across multiple *queries* increases the quality of the answers, a *feature* called "self-consistency."

This is particularly important when LLMs have access to tools or databases. Based on the *queries*, one can decide to use tools; for example, we can provide examples of question-action pairs: "What is the age of the universe?" -> [search Wikipedia]).

Usually, question types can be divided into *zero-shot prompts* and *few-shot prompts*, as shown in the diagram in Figure 5.1.

This is because people usually ask direct questions to the model (zero-shot), or examples are provided to the model (*Few shots*). This takes the form of *prompts* like the following:

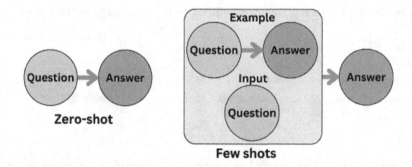

FIGURE 5.1   Zero-shot and Few-shot prompts.

```
"""
Sample question
Sample answer

Question
What is the answer?
"""
```

In *few-shot prompts*, you can get better results by showing the reasoning leading to a specific answer, i.e., your CoT, and you can induce *zero-shot*-like behavior by asking it to "think step-by-step," known as *inception*. The CoT can take the form of intermediate questions and answers (i.e., self-ask) and can induce specific answers by referring to concepts or analogies. If you re-execute the same *query* several times, you will get different answers, so choosing a consistent answer over several *queries* increases the quality of the answers, which is called *self-consistency*.

The above allows chaining *prompts* to perform complex tasks that must be performed step by step, perhaps following a plan. This can solve complex problems by inducing a plan of action (see Figure 5.2). Each step of the plan can be used to generate its own chain of actions.

Thus, LLMs can be thought of as a kind of flexible subroutines that take inputs and produce outputs. *Prompts* are the molds that shape the subroutines to solve a specific problem. LLM applications consist of assembling these subroutines to create novel capabilities.

An example of CoT prompts includes:

 You

The odd number in this group sum up an even number: 4, 8, 9, 15, 12, 2, 1. A: by suming up the odd numbers (9, 15, 1), the number 26 is obtained. The answer is false. The odd numbers of this group sum up an even number: 15, 32, 5, 13, 82, 7, 1. A:

 ChatGPT

Let's evaluate the sum of the odd numbers in the second group (15, 5, 13, 7, 1):

$$15 + 5 + 13 + 7 + 1 = 41$$

The sum of these odd numbers is 41, which is an odd number. Therefore, the statement "The odd numbers of this group sum up an even number" is false, not true. If you have any further questions or if there's anything else you'd like to discuss, feel free to let me know!

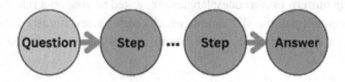

FIGURE 5.2 Chaining of prompts.

5.10 CHATGPT CONVERSATIONAL SYSTEM

ChatGPT[1] is an NLP tool created by the company OpenAI that enables human-like conversations and much more with a *chatbot*-like conversational system (Shen et al., 2023). The system can answer questions or follow instructions to perform various types of tasks, such as writing texts, summarizing documents, answering questions, generating code, translating texts, etc. (Hariri, 2023).

The tool can be accessed interactively or through APIs in programming languages such as Python.

ChatGPT runs using the GPT model in its GPT-3.5 and GPT-4 versions. Unlike other models, GPT-3 and later use the RLHF technique, in order to allow human feedback in the initial training of the model. Through RLHF, human trainers provided the model with conversations in which both the user and the AI assistants played a role. RLHF reward methods help determine the best responses. To further train the *chatbot*, users could vote for or against their answer by clicking on the "thumbs up" or "thumbs down" icons next to the answer. To collect this data, conversations that AI coaches had with the *chatbot* were taken. A message written by a model was randomly selected, several alternative endings were tested, and the AI trainers were asked to rank them. Using these reward models (RMs), the initial pre-trained model (GPT-3-5 or GPT-4) can be adjusted.

Users can ask ChatGPT a variety of questions or instructions called *prompts*. These include simple or more complex questions such as "What is the meaning of life?" or "In what year was Santiago founded?". Given the training *datasets* used in GPT3-5 and GPT-4, ChatGPT is proficient in scientific disciplines and can debug or write code. There is no limitation on the types of questions to ask ChatGPT. However, ChatGPT uses data up to the year 2021, so it has no knowledge of events and data after that year. And, since it is a conversational *chatbot*, users can request more information or ask it to try again when generating text.

ChatGPT has been trained using the same methods as pre-trained instruction-based models (i.e., InstructGPT), but with slight differences in the data collection configuration.

ChatGPT uses the RLHF method consisting of three steps:

1. **Supervised *fine-tuning***: a pre-trained LLM is fine-tuned on a relatively small amount of demo data selected by labelers, to learn a supervised policy (the SFT model), which generates results from a selected list of *prompts*. This represents the reference model.

2. **Mimicking human preferences**: labelers are asked to vote on a relatively large number of outcomes from the SFT model, which creates a new *dataset* consisting of comparison data. A new model is trained on this *dataset*. This is referred to as a "reward model."

3. **Proximal policy optimization (PPO)**: the RM is used to further refine and improve the SFT model. The result of this step is the so-called "policy model."

Step 1 is performed only once, while steps 2 and 3 can be repeated continuously: more comparison data is collected on the current best policy model, which is used to train a new RM and then a new policy.

Step 1: The supervised *fine-tuning* model

In this step, demonstration data are collected to train a supervised policy model, called the "SFT model":

- *Data collection*: a list of *prompts* is selected, and a group of human labelers are asked to write the expected output response. Since this process is slow and expensive, the result is a relatively small, high- quality *dataset* that is used to fit a pre-trained LLM.

- *Model choice*: instead of tuning the original GPT-3 model, a pre-trained GPT-3.5 model based on a reference model called "text-davinci-003" was used.

    Due to the limited amount of data for this step, the obtained SFT model could generate text that is still misaligned. To overcome this problem, labelers should classify different outputs of the SFT model to create a RM.

Step 2: The RM

The goal is to learn an objective function (i.e., RM) directly from the data, giving a score to the outputs of the SFT model, proportional to how desirable these outputs are to humans. In the end, this process will extract from the data an automatic system that is supposed to mimic human preferences.

To achieve the above, a list of requests is selected and the SFT model generates multiple outputs (between four and nine) for each request. The labelers rank the outputs from best to worst. The result is a newly labeled *dataset*, which is used to train a RM, which takes as input some of the outputs of the SFT model and ranks them in order of preference.

In practice, this *dataset* has been generated from a selection of thirty to forty thousand *prompts*, and a variable number of the generated outputs (for each *prompt*) are presented to each tagger during the ranking phase.

Step 3: *Fine-tuning* of the SFT model through PPO

Reinforcement learning is now applied to fine-tune the SFT policy, allowing it to optimize the RM. The specific algorithm used to train the agent by reinforcement is called *proximal policy optimization* (PPO).

It is called a "policy algorithm" because it learns and updates the current policy directly, rather than from past experience as in "out-of-policy" algorithms such as Deep Q-Network. This means that PPO is continuously adapting the current policy based on the actions the agent performs and the rewards it receives.

PPO uses a trust region optimization method to train the policy, which means that it restricts the change in the policy to be within a certain distance from the previous policy to ensure stability. The method uses a value function to estimate the expected return of a given state or action, which is used to calculate the advantage function, representing the difference between the expected return and the actual return. Subsequently, the function is used to update the policy by comparing the action taken by the current policy with the action that would have been taken by the previous policy.

In this step, the PPO model is initialized from the SFT model, and the value function is initialized from the RM. The environment is a bandit environment that presents a random *prompt* and expects a response to the *prompt*. Given the *prompt* and the response, it produces a reward (determined by the RM) and the episode ends. A Kullback-Leibler divergence penalty per *token* of the SFT model is added to each *token* to mitigate over-optimization of the RM.

5.10.1 Performance Evaluation

Because the model is trained on the input of human labelers, the central part of the evaluation is also based on human input; that is, it is performed by having the labelers rate the quality of the model outputs. To avoid overfitting to the judgment of the labelers involved in the training phase, the test set uses cues from retained OpenAI clients that are not represented in the training data.

The model can be evaluated according to three high-level criteria:

1. **Usability**: it judges the model's ability to follow user instructions as well as to infer instructions.

2. **Truthfulness**: it judges the tendency of the model to hallucinate (make up facts) in closed domain tasks. The model is evaluated on the TruthfulQA *dataset*.

3. **Harmlessness**: it evaluates whether the model output is appropriate, denigrates a protected class, or includes derogatory content. The model is also compared to the RealToxicity*Prompts* and CrowS-Pairs *datasets*.

The model is also evaluated to determine *zero-shot* performance on traditional NLP tasks such as question answering, reading comprehension, and summarization.

In practice, ChatGPT can usually be interacted with through various examples of *prompts*. However, for application development purposes, you can use the APIs, which allow you to access the models for various types of tasks.

In this example we will see two simple cases: a simple application of *prompts* for sentence *completion* and another case for *following instructions*.

1. ***Prompt* Completion**

For this example, you should use the program "`Example-ChatGPT-1`". This is a basic use case of the OpenAI API for accessing pre-trained GPT-based models. The different varieties of models can be 3-5 (turbo). The task consists of performing a *prompt* on a question.

The API access key must be obtained directly from OpenAI.

First, we install some packages:

```
!pip install openai
```

We import some libraries:

```
!pip install openai
```

We set our API access key:

```
openai.api_key = "INSERT API KEY"
```

Then, we invoke the API to perform the chat "completion" task through a question *prompt*. For this, we specify the model to use ("*gpt-3.5-turbo*") and the roles in the interaction. For this, we will assume two basic roles:

- *System*: it specifies the role to be taken by ChatGPT (e.g., a wizard).

- *User*: it specifies the role of the user asking the question.

```
response = openai.ChatCompletion.create(
    model ="gpt-3.5-turbo",
    messages=[
        {"role": "system", "content": "Assistant is an LLM trained
by OpenAI."},
        {"role": "user", "content": "Who founded Santiago de
Chile?"}
    ]
)
```

Finally, we get the answer (usually just an option or choice about the content):

```
print(response['choices'][0]['message']['content'])
```

```
The city of Santiago de Chile was founded on February 12, 1541 by
the Spanish conquistador Pedro de Valdivia. Valdivia led the
expedition that conquered the territory and established the city,
making it the capital of the Kingdom of Chile.
```

2. Instruction Following

For this example, we load the program in Colab "Example-ChatGPT-2". This is a basic example of instruction tracking at a *prompt* using the OpenAI API to access pre-trained GPT-based models. The different varieties of models can be 3–5 (turbo). The task is to perform a *prompt* based on an instruction.

The API access key must be obtained directly from OpenAI.

First, we install some packages:

```
!pip install openai
```

Then, we import some libraries to use the GPT models:

```
import openai
```

We set our API access key:

```
openai.api_key = "INSERT API KEY"
```

We load a sample text:

```
MyText = open("./sample.txt").read()
```

We prepare a *prompt* to request to generate a summary of the input text:

```
MyPrompt = "Make a summary of the following text: "+ MyText
```

Then, we invoke the API to perform an instruction *prompt*; in this case, summarize an input text, based on the pre-trained model ("*gpt-3.5-turbo*"), with a maximum number of output tokens, and a temperature parameter that regulates the randomness of the response that is generated:

```
response = openai.ChatCompletion.create(
    model   ="gpt-3.5-turbo",
    messages=[{"role": "user", "content": MyPrompt}],
    max_tokens=500,
    temperature=0.7,
)
```

It displays the generated response:

```
print(response['choices'][0] ['message']['content'])
```

```
The text informs that the General Director of Carabineros, Bruno
Villalobos Krumm, has resigned from his post, and mentions that
the Government is committed to modernizing Carabineros and the
Investigative Police, as well as improving coordination between
police, prosecutors and judges and strengthening the capacity to
rehabilitate inmates. The former Director General is also thanked,
and it is announced that the current Deputy Director General,
Julio Pineda Peña, will assume command of Carabineros as deputy.
```

5.11 BARD CONVERSATIONAL SYSTEM

BARD[2] is a *chatbot* based on the PalM-2 model developed by Google AI and trained with a massive dataset of text and code. Similar to ChatGPT, it can generate text, translate languages, write different types of creative content, and answer your questions in an informative way.

Unlike other conversational systems, BARD has the following advantages:

- **Multilingualism**: PaLM-2 is further trained in multilingual text, covering more than one hundred languages. This significantly improves its ability to understand, generate and translate nuanced text, including idioms, poems and riddles, in a wide variety of languages, a difficult problem to solve.

- **Reasoning**: The PaLM-2 dataset includes scientific articles and web pages containing mathematical expressions. As a result, it can demonstrate enhanced capabilities in logic, common sense reasoning and mathematics.

- **Coding**: PaLM-2 is pre-trained on a large dataset of publicly available source code. This means that it excels in popular programming languages such as Python and JavaScript but can also generate specialized code in languages such as Prolog, Fortran, and Verilog.

In addition, BARD through PaLM-2 brings advanced AI capabilities directly into several Google products including:

- PaLM-2's enhanced multilingual capabilities allow BARD to expand into new languages.

- Workspace *feature* s to help you write in Gmail and Google Docs, and help you organize in Google Spreadsheets.

On the other hand, there are some relevant differences between BARD and ChatGPT, as shown in Table 5.1.

TABLE 5.1   Comparison between ChatGPT and BARD

Characteristic	ChatGPT	BARD
Language	GPT 3-5 or GPT-4	LaMDA y PaLM-2
Type of model	*Offline* model with data available until 2021 (no real-time data)	Online model taken from the web directly
Learning algorithm	RLHF	*Transformers* trained on dialogue
Integration	Various applications, such as Bing, Duolingo or Snapchat	Google search engine
Answers	Can only create one answer	Can create three answers that the user can select
Web search	Answers based on training data only	Can search the web and deliver the references to the user
Classifier	GPT-4 has its own classifier, which can detect self-generated text	BARD has no classifiers yet

5.12 CONCLUSIONS

In this chapter, we have explored a variety of practical exercises that can be performed with LLMs, revealing their versatility and power in various applications. Text classification has proven to be one of the most common and effective tasks for these models, as they can analyze and understand complex contexts to assign accurate labels to documents. Fine-tuning, on the other hand, has opened new perspectives for model customization, allowing them to be tailored to specific tasks and domains.

Semantic search has emerged as an intriguing and valuable exercise that capitalizes on the ability of large language models to understand the meaning and similarity between words and phrases. This has led to advances in question-answer processing, information retrieval, and improved recommender systems. Models have also excelled in causal inference, opening up possibilities in assessing the impact of certain events or decisions in different scenarios.

Finally, we have explored proprietary database access with our LLMs and the design of *prompts* that can guide large language models toward specific tasks. These structured *prompts* allow us to harness the power of the language to perform tasks such as text generation, software design, and creative content creation.

NOTES

1 https://chat.openai.com/
2 https://bard.google.com/

## CHAPTER 6

Issues and Perspectives

6.1 INTRODUCTION

Large Language Models (LLMs) have the potential to have a dramatic effect on businesses and jobs alike. Although LLMs have been around for some time, Pandora's box has recently been opened by providing access that allows businesses and organizations to use the power of artificial intelligence (AI) to automate jobs. This can have both positive and negative effects.

This chapter explores various considerations related to the present and future of LLMs, including emerging skills, alignment, ethics, regulations, risks, benefits, and limitations.

6.2 EMERGING SKILLS

LLMs pose a significant concern because of their tendency to exhibit *emerging risk behaviors* (Boiko, MacKnight, and Gomes, 2023). These behaviors may include formulating protracted plans, pursuing undefined goals, and striving to acquire additional authority or resources. Due to the complex nature of LLMs, it is not easy to predict how they will behave in specific situations.

Emergence can be defined as the sudden appearance of novel behavior. Apparently, LLMs show emergence by suddenly acquiring new skills as they grow. Why does this happen and what does it mean?

In recent years, significant efforts have been made to scale LLMs, leading to steady and predictable improvements in their ability to learn these patterns, which can be seen in improvements in quantitative metrics.

In addition, the scaling process leads to an interesting qualitative behavior: as LLMs are scaled, a number of critical scales are reached where, suddenly, new skills are "unlocked." LLMs are not directly trained to have these skills and they appear quickly and unpredictably, as if emerging out of nowhere. These emergent skills include performing arithmetic, answering questions, summarizing passages, and more, which LLMs learn simply by observing natural language.

DOI: 10.1201/9781003517245-6

6.2.1 What Causes These Emergent Skills and What Do They Mean?

While the fact that LLMs gain these skills as they scale is remarkable, the way they emerge is especially interesting. In particular, many LLM skills seem to emerge; that is, as LLMs grow in size, they go from near-zero performance to, at times, state-of-the-art performance at an incredibly rapid pace and on unpredictable scales.

Emergent behavior is not unique to LLMs and, in fact, is seen in many fields, such as physics, evolutionary biology, economics, and dynamical systems. In general, emergence is an essential phenomenon of small changes in the quantitative parameters of a system causing large changes in its qualitative behavior. The qualitative behavior of these systems can be viewed as different "regimes," in which the "rules of the game," or equations that dictate behavior, can vary dramatically.

The scaling of LLMs has shown consistent and predictable improvements in performance, with the scaling law for the cross-entropy loss of LLMs holding over seven orders of magnitude.

The important part of this phenomenon lies in the fact that we, *a priori*, do not know in advance that this will happen or even at what scale it might happen. Thus, while we can try to come up with new architectures or some other novel invention to address complicated natural language problems, we can solve these problems by simply scaling LLMs to be even larger.

Even if something simple like multistep reasoning is an important explanatory factor for emergent abilities, its mere existence is still important. Ultimately, if completing the tasks that really matter to us humans requires multistep reasoning, and it is likely that many of them do, then it really doesn't matter if there is a simple explanation for emergent abilities. The simple observation that scaling models can increase their performance in real-world applications is sufficient.

6.3 LLM IN PRODUCTION

In *prompt* design, instructions are written in natural languages, which are much more flexible than programming languages. This can not only generate an excellent user experience but can also lead to a rather poor developer experience. Flexibility comes from two directions: how users define *prompts* and how LLMs respond to these *prompts*.

Flexibility in user-defined *prompts* leads to silent failures. If someone accidentally makes some changes to the code, such as adding a random character or deleting a line, it is likely to throw an error. However, if someone accidentally changes a *prompt*, it will still run, but it will give very different results. While the flexibility in user-defined *prompts* is only a nuisance, the ambiguity in the responses generated by LLMs can be a deciding factor. It leads to two problems:

1. **Ambiguous output format**: downstream applications, in addition to LLMs, expect outputs in a particular format so that they can be parsed. We can design our *prompts* to be explicit about the output format, but there is no guarantee that outputs will always follow this format.

2. **Inconsistency in the user experience**: When using an application, users expect some consistency. Imagine an insurance company that offers you a different quote every time you visit their website. LLMs are stochastic: there is no guarantee that an LLM will give you the same result for the same input every time.

The above may force an LLM to give the same answer by setting *temperature*=0, which is generally a good practice. While this mainly solves the consistency problem, it does not inspire confidence in the system. Imagine a teacher giving you consistent scores only if that teacher sits in a particular room. If that teacher sits in different rooms, that teacher's scores for you will be unbelievable.

On the other hand, the more explicit details, and examples you include in the *prompt*, the better the performance of the model and the more expensive it is to infer; for example, the OpenAI API charges for input and output tokens. Depending on the task, a simple *prompt* can have between three hundred and one thousand tokens. If you want to include more context, this can easily reach 10k tokens just for the *prompt*.

Another very promising direction is to use LLMs to generate *embeddings* and then build applications on top of these *embeddings*, e.g., for a search problem. As of April 2023, the cost of *embeddings* with the smallest OpenAI model is $0.0004/1k tokens. If each article averages 250 tokens (187 words), this price means $1 per 10k articles or $100 per 1 million articles.

The primary cost of incorporating models for real-time use cases is to load these *embeddings* into a vector database for low-latency retrieval. It is exciting to see so many new vector databases flourishing, such as Pinecone, Qdrant, Weaviate, or Chroma, as well as the existing Faiss, Redis, Milvus, or ScaNNN.

6.4 HUMAN-LLM ALIGNMENT

The capability of a model is generally evaluated by how well it can optimize its objective function, the mathematical expression that defines the model's objective; for example, an LLM designed to predict stock market prices may have an objective function that measures the accuracy of the model's predictions. If the model is able to accurately predict the movement of stock prices over time, it would be considered to have a high level of capability for this task.

On the other hand, alignment has to do with what we actually want the model to do versus what it is being trained to do. It asks the question, "Is such a target function consistent with our intentions?" And it refers to the extent to which a model's goals and behavior align with human values and expectations. For a simple concrete example, suppose we train a bird classifier to classify birds as "sparrows" or "robins" and use logarithmic loss (which measures the difference between the model's predicted probability distribution and the actual distribution) as the target training, even though our ultimate goal is high classification accuracy. The model may have low log loss, i.e., model capability is high, but low accuracy on the test data. In fact, the log loss is not perfectly correlated with the accuracy of classification tasks. This is an example of misalignment, where the model is capable of optimizing the training objective, but misaligned with our final objective; for example, models like GPT-3 are misaligned.

LLMs like GPT-3 are trained on large amounts of text data from the Internet and are able to generate human-like text, but they may not always produce results consistent with human expectations or desirable values. In fact, their objective function is a probability distribution over sequences of words (or sequences of tokens), which allows them to predict what the next word in a sequence is (more details on this below).

However, in practical applications, these models are intended to do some valuable cognitive work, and there is a clear divergence between how these models are trained and how we would like to use them. Although a machine-calculated statistical distribution of word sequences might be, mathematically speaking, a very efficient choice for modeling language, we, as humans, generate language by choosing the sequences of text that are best for the given situation, using our prior knowledge and common sense to guide us in this process. This can be a problem when language models are used in applications that require a high degree of trust or reliability, such as dialog systems or intelligent personal assistants.

While these powerful and complex models trained on large amounts of data have become extremely capable in recent years, when used in production systems to facilitate human life, they often fall short of this potential. The alignment problem in LLMs usually manifests itself as follows:

- **Absence of help**: not following the user's explicit instructions.

- **Hallucinations**: model inventing non-existent or erroneous facts.

- **Lack of interpretability**: it is difficult for humans to understand how the model arrived at a certain decision or prediction.

- **Generation of biased or toxic results**: a language model that is trained with biased/toxic data may reproduce it in its output, even if it was not explicitly instructed to do so.

6.5 ETHICS

As LLMs become more powerful, it is vital to consider the ethical implications of their use. From the generation of harmful content to the disruption of privacy and the spread of misinformation, the ethical concerns surrounding the use of LLMs are complicated and manifold:

- **Generation of harmful content**: LLMs have the potential to generate harmful content, such as hate speech, extremist propaganda, racist or sexist language, and other forms of content, which could cause harm to specific individuals or groups. While the models are not inherently biased or harmful, the data they are trained on may reflect biases that already exist in society. This, in turn, can lead to serious social problems, such as incitement to violence or increased social unrest.

- **Economic impact**: LLMs can also have a significant economic impact, particularly as they become increasingly powerful, widespread, and affordable. They may introduce substantial structural changes in the nature of work and labor, such as making certain jobs redundant through the introduction of automation. This could result in workforce displacement, mass unemployment, and exacerbate existing inequalities in the workforce.

- **Hallucinations**: a major ethical concern related to LLMs is their tendency to hallucinate, i.e., to produce false or misleading information using their internal patterns and biases. While some degree of hallucination is inevitable in any language model, the extent to which it occurs can be problematic. This can be especially damaging as models become increasingly convincing and users without domain-specific knowledge begin to rely too heavily on them. It can have serious consequences for the accuracy and veracity of the information generated by these models. Therefore, it is essential to ensure that AI systems are trained on accurate and contextually relevant datasets to reduce the incidence of hallucinations.

- **Privacy**: LLMs also raise important questions about user privacy. These models require access to large amounts of data for training, which often includes people's personal data. This is generally collected from licensed or publicly available *datasets* and can be used for various purposes, such as finding geographic locations based on phone codes available in the data.

- **Biases and ethical concerns**: LLMs may reinforce biases present in their training data. This could raise ethical concerns, such as discriminatory or misleading content that tarnishes a company's reputation.

6.6 REGULATORY ISSUES

The accountability of responses generated by LLMs is an important point for possible regulatory action and can directly influence their use and improve the quality of responses. Initially, platforms disclaim responsibility for the responses generated by their models, holding their users accountable. The models are trained with a huge amount of data available on the Internet and texts that are eventually part of some published or copyrighted scientific work. In this sense, the regulation should be very clear about who is responsible for the use of the answers since, for example, they may be the result of plagiarism.

Measuring or mitigating bias and discrimination in LLMs is a complex task, considering that LLMs are trained with a large dataset that can reproduce the biases observed in society. Thus, the regulation of LLMs can be done through a data representativeness audit process, i.e., assessing whether the data used to train the model are representative of the diversity observed in society (as in the case of credit risk rating systems or facial recognition systems). In addition, models can be evaluated through bias metrics, such as equalized probabilities (measuring whether false positive and false negative rates are equal across different groups, such as genders or races). Another applicable strategy is human evaluation. Using this type of strategy, model results should be exposed to human evaluators to determine whether the results generate any bias or discrimination (this is a costly method but can be used to observe discriminatory nuances in the results).

6.7 COMPLEXITY

In simple terms, a *parameter* is a component of a machine learning model that the model learns from its training data; for example, in a neural network, parameters include the weights and biases of each node in the network, which determine how each node processes its input, and adjusting these values allows the model to learn patterns in the data.

The number of parameters in an LLM significantly affects the fitting process in several ways:

- **Model complexity and learnability**: a model with more parameters can represent more complex patterns and has a higher learnability. This means that it can potentially achieve better performance when fitting on a specific task, as it can learn more nuanced representations of the data. However, it also means that the model may require more fitting data to avoid overfitting.

- **Computational resources:** models with more parameters require more computational resources (memory and processing power) for fine-tuning. This can make the fitting process more challenging and time-consuming, especially for those who do not have access to high-end hardware or the capital for hosting services.

- **Risk of overfitting**: a model with more parameters can learn more complex patterns, but it also has a higher risk of overfitting, especially if the fine-tuning data is small. Overfitting occurs when the model learns the training data too well, including noise and outliers, and performs poorly on unseen data.

- **Transfer learning:** models with more parameters often perform better in transfer learning, where the model is first trained on a large data set (pre-training) and then fit on a smaller, task-specific data set. The large number of parameters allows the model to learn a wide range of language patterns during pre-training, which can then be tuned for specific tasks.

6.8 RISKS

One of the main implications of LLMs is their potential to increase automation and improve efficiency in various industries. LLMs can be used to automate various tasks, such as customer service, data entry, and content creation, which can lead to cost savings and increased productivity; for example, GPT-3 can be used to generate natural language text with impressive consistency and creativity, which could enable the creation of high-quality content at scale.

However, with the increasing use of LLM, there are also concerns about potential job losses as automation replaces human labor in a number of industries. It is also possible that the use of LLM could lead to biases in decision making, as these models learn from the data they are fed and, if that data is biased, it can lead to biased decisions. This is particularly relevant in industries such as finance, where biased models could lead to unfair lending practices.

Another concern is the potential for malicious actors to use LLM for nefarious purposes, such as generating fake news and propaganda, phishing attacks or even *phishing*. The same natural language generation capabilities that make GPT-3 impressive could be used to spread misinformation and propaganda.

6.9 LIMITATIONS

There are several limitations to LLMs that, in the coming years, should be addressed by the international scientific community:

- **Sophisticated but probabilistic autocompletion machines**: LLMs are essentially an autocompletion machine that operates using sophisticated pattern recognition methods. It repeats and reconstructs the prose in which it has been trained, but it does so probabilistically; thus, it will often make mistakes and even invent "facts" or produce invented references. These "hallucinations" are not an aberration; they are an inherent behavior of any generative system.

- **Text generation skills are not suitable in contexts with low fault tolerance**: the corollary of the first limitation is that LLM text generation skills are less suitable for tasks with low "fault tolerance"; for example, legal writing and tax advice have low fault tolerance and are very specific use cases that require expertise, accountability, and confidence, not just words on pages.

- **Creative writing tasks may have high fault tolerance**: the wide variation in LLMs may actually be a *feature* rather than a bug, as it can help generate creative and valuable options that humans had not considered before. Even human moderation is necessary to avoid exposing users to uninhibited and potentially mind-boggling *chatbot* responses.

- **Privacy and data security risks**: there are likely to be legal and regulatory restrictions on the movement and storage of your organization's and your customers' data. The widespread use of ChatGPT among knowledge workers presents a serious risk of exposing sensitive data to a third-party system. This data will be used to train future versions of OpenAI models, which means it could potentially be regurgitated to the public.

6.10 CONCLUSIONS

LLMs exhibit emergent risk behaviors and exhibit skills that appear unpredictably as they scale. These emergent abilities may be beneficial in terms of steady improvements in model performance, but they also raise ethical concerns, such as the generation of harmful content, a lack of consistency in results, and the possibility of hallucinations or the generation of false information. In addition, LLMs can have economic implications, such as the automation of jobs and the impact on the workforce. It is essential to consider the risks and ethical implications of LLMs and find ways to regulate their use to ensure responsible and beneficial employment.

In the future, efforts are likely to be made to address the ethical concerns and risks associated with LLMs. This may include implementing stricter regulations and standards to ensure alignment with human values and reduce bias in the results generated. Advances are also expected to be made in the ability to interpret and understand LLM decisions and predictions, which may help address the lack of interpretability. In addition, improvements are expected to be made in the privacy and security of data used by LLMs, with a focus on protecting users' confidential information. Overall, LLMs are expected to continue to play an important role in automating language-related tasks, but with an increased focus on ensuring responsible and ethical use of this technology.

Bibliography

Achille, A., & Soatto, S. (2017). *On the Emergence of Invariance and Disentangling in Deep Representations.*arXiv, abs/1706.01350.

Adiwardana, D., Luong, M.-T., So, D. R., Hall, J., Fiedel, N., Thoppilan, R., ... Le, Q. V. (2020). *Towards a Human-like Open-Domain Chatbot.* arXiv:2001.09977.

Agarwal, B., Nayak, R., Mittal, N., & Patnaik, S. (2020). *Deep Learning-Based Approaches for Sentiment Analysis.* Springer.

Aggarwal, C. (2018). *Machine Learning for Text.* Springer.

Alto, V. (2023). *Modern Generative AI with ChatGPT and OpenAI Models: Leverage the Capabilities of OpenAI's LLM for Productivity and Innovation with GPT3 and GPT4.* Paxckt.

Atkinson, J. (2005). Intelligent search agents using web-driven natural-language explanatory dialogs. *IEEE Computer, 38,* 44–52.

Atkinson, J. (2022). *Text Analytics: An Introduction to the Science and Applications of Unstructured Information Analysis.* Taylor & Francis, CRC Press.

Atkinson, J., & Palma, D. (2018). Coherence-based automatic essay assessment. *IEEE Intelligent Systems, 33*(5), 26–36.

Atkinson-Abutridy, J., Mellish, C., & Aitken, S. (2003). A semantically guided and domain-independent evolutionary model for knowledge discovery from texts. *IEEE Transactions on Evolutionary Computation, 7,* 546–560.

Babcock, J., & Bali, R. (2021). *Generative AI with Python and TensorFlow 2: Create Images, Text, and Music with VAEs, GANs, LSTMs, Transformer Models.* Packt Publishing.

Baron, M. (2019). *Probability and Statistics for Computer Scientists.* Chapman and Hall/CRC.

Beck, S. C. (1988). Improving information retrieval using latent semantic indexing. *Proceedings of the ACM SIGOIS and IEEE CS TC-OA COIS90: conference on Office information systems,* Cambridge Massachusetts, USA. 40–47.

Bengio, I. J.-A.-F. (2014). Generative Adversarial Networks. arXiv:1406.2661.

Berant, G. T. (2019). *Evaluating Text GANs as Language Models.* arXiv:1810.12686.

Bermúdez, J. (2020). *Cognitive Science: An Introduction to the Science of the Mind.* Cambridge University Press.

Bernstein, J. S. (2023). *Generative Agents: Interactive Simulacra of Human Behavior.* arXiv:2304.03442.

Bohnet, B., Mcdonald, R., Andor, D., Pitler, E., & Maynez, J. (2018). *Morphosyntactic Tagging with a Meta-BiLSTM Model over Context Sensitive Token Encodings.* Melbourne: Association for Computational Linguistics (ACL)..

Boiko, D. A., MacKnight, R., & Gomes, G. (2023). *Emergent Autonomous Scientific Research Capabilities of Large Language Models.* arXiv:2304.05332.

Bokka, K., Hora, S., & Jain, T. (2019). *Deep Learning for Natural Language Processing : Solve your Natural Language Processing Problems with Smart Deep Neural Networks.* Packt Publishing.

Bommasani, R., Hudson, D. A., Adeli, E., Altman, R. B., Arora, S., von Arx, S., ... Liang, P.(2021). *On the Opportunities and Risks of Foundation Models.*arXiv, abs/2108.07258.

Brown, T. B., Mann, B., Ryder, N., Subbiah, M., Kaplan, J., Dhariwal, P., ... Amodei, D.(2020). *Language Models are Few-Shot Learners.*arXiv, abs/2005.14165.

Burns, S. (2019). *Natural Language Processing : A Quick Introduction to NLP with Python and NLTK.* Amazon.com Services.

Choi, R. Z. (2019). HellaSwag: Can a Machine Really Finish Your Sentence? arXiv:1905.07830.

Chowdhery, A., Narang, S., Devlin, J., Bosma, M., Mishra, G., Roberts, A., …, Fiedel, N. (2022). *PaLM: Scaling Language Modeling with Pathways.* arXiv:2204.02311.

Christiano, P., Leike, J., Brown, T. B., Martic, M., Legg, S., & Amodei, D. (2023). *Deep Reinforcement Learning from human preferences.* arXiv:1706.03741.

Cuantum, T. (2023). *Introduction to Natural Language Processing with Transformers: Decoding Language with AI: A Comprehensive Guide to Build Language Applications with Hugging … Python, and More.* Cuantum Technologies.

Dai, D., Sun, Y., Dong, L., Hao, Y., Sui, Z., & Wei, F. (2022). *Why Can GPT Learn In-Context? Language Models Secretly Perform Gradient Descent as Meta-Optimizers.* arXiv:2212.10559.

Dai, Z., Yang, Z., Yang, Y., Carbonell, J. G., Le, Q. V., & Salakhutdinov, R. (2019). *Transformer-XL: Attentive Language Models beyond a Fixed-Length Context.* arXiv, abs/1901.02860.

Darling, K. (2022). *ChatGPT: A Scientist Explains the Hidden Genius and Pitfalls of OpenAI XE "OpenAI" 's XE "OpenAI's" chatbot.* Science Focus.

de Santana Correia, A., & Colombini, E. L. (2021). *Attention, Please! A Survey of Neural Attention Models in Deep Learning.* arXiv, abs/2103.16775.

Deng, L., & Liu, Y. (2018). *Deep Learning in Natural Language Processing.* Springer.

Devlin, J., Wei, M., Kenton, C., & Toutanova, L. (2019). BERT: Pre-training of Deep Bidirectional *Transformers* for Language Understanding. *Proceedings of NAACL HLT 2019,* North American Chapter of the Association for Computational Linguistics, 4171–4186.

Eisenstein, J. (2019). *Introduction to Natural Language Processing.* The MIT Press.

Ekman, M. (2022). *Learning Deep Learning: Theory and Practice of Neural Networks, Computer Vision, Natural Language Processing, and Transformers Using TensorFlow.* Addison-Wesley.

Foster, D. (2019). *Generative Deep Learning: Teaching Machines to Paint, Write, Compose and Play.* O'Reilly.

Freitag, M., & Al-Onaizan, Y. (2017). Beam Search Strategies for Neural Machine Translation. In *Proceedings of the First Workshop on Neural Machine Translation,* Association for Computational Linguistics, Vancouver, 56–60.

Gao, S., & Kean, A. (2023). *On the Origin of LLMs: An Evolutionary Tree and Graph for 15,821 Large Language Models.* arXiv:2307.09793.

Ge, Y., Hua, W., Ji, J., Tan, J., Xu, S., & Zhang, Y. (2023). *OpenAGI: When LLM Meets Domain Experts.* arXiv:2304.04370.

Ghosh, S., & Gunning, D. (2019). *Natural Language Processing Fundamentals: Build intelligent applications that can interpret the human language to deliver impactful results.* Packt Publishing.

Gillon, B. (2019). *Natural Language Semantics: Formation and Valuation.* The MIT Press.

Goldberg, Y. (2017). *Neural Network Methods for Natural Language Processing.* Morgan & Claypool Publishers.

Gooding, S., & Kochmar, E. (2019). Complex word identification as a sequence labelling task. *Proceedings of the 57th Annual Meeting of the Association for Computational Linguistics* (pp. 1148–1153). Florence: Association for Computational Linguistics.

Graves, A. (2012). *Supervised Sequence Labelling with Recurrent Neural Networks.* Springer.

Guilla, H. T.-A. (2023). *LLaMA: Open and Efficient Foundation Language Models.* arXiv:abs/2302.13971.

Hariri, W. (2023). *Unlocking the Potential of ChatGPT: A Comprehensive Exploration of its Applications, Advantages, Limitations, and Future Directions in Natural Language Processing.* arXiv:2304.02017.

Howard, J., & Ruder, S. (2018). *Fine-tuned Language Models for Text Classification.* arXiv, abs/1801.06146.

Hu, Z., Lan, Y., Wang, L., Xu, W., Lim, E.-P., Lee, R. K.-W., … Poria, S. (2023). *LLM-Adapters: An Adapter Family for Parameter-Efficient Fine-Tuning of Large Language Models.* arXiv:2304.01933.

Huang, T., Hsieh, C., & Wang, H. (2018). Automatic meeting summarization and topic detection system. *Data Technologies and Applications*, 351–365.

Kalyan, K. S., Rajasekharan, A., & Sangeetha, S. (2021). *AMMUS: A Survey of Transformer-Based Pretrained Models in Natural Language Processing.* arXiv:2105.00827.

Kamath, U., Liu, J., & Whitaker, J. (2019). *Deep Learning for NLP and Speech Recognition.* Springer.

Kendall, E., & McGuinness, D. (2019). *Ontology Engineering.* Morgan & Claypool Publishers.

Kublik, S., & Saboo, S. (2022). *GPT-3.* O'Reilly Media.

Kudo, T., & Richardson, J. (2018). SentencePiece: A simple and language independent subword tokenizer and detokenizer for Neural Text Processing. *Proceedings of the 2018 Conference on Empirical Methods in Natural Language Processing: System Demonstrations*, Association for Computational Linguistics, Brussels, Belgium, 66–71.

Kuhn, M., & Johnson, K. (2019). *Feature Engineering and Selection: A Practical Approach for Predictive Models.* Chapman and Hall/CRC.

Lane, H., Hapke, H., & Howard, C. (2019). *Natural Language Processing in Action: Understanding, analyzing, and generating text with Python.* Manning Publications.

Lewis, M., Liu, Y., Goyal, N., Ghazvininejad, M., Mohamed, A., Levy, O., … Zettlemoyer, L. (2019). *BART: Denoising Sequence-to-Sequence Pre-training for Natural Language Generation, Translation, and Comprehension.* arXiv, abs/1910.13461.

Li, H. (2022). Language models: past, present, and future. *Communications of the ACM, 65*, 56–63.

Lialin, V., Deshpande, V., & Rumshisky, A. (2023). *Scaling Down to Scale Up: A Guide to Parameter-Efficient Fine-Tuning.* arXiv.

Liang, P. P., Wu, C., Morency, L.-P., & Salakhutdinov, R. (2021). *Towards Understanding and Mitigating Social Biases in Language Models.* arXiv:2106.13219.

Liu, F., & Perez, J. (2017). Gated End-to-End Memory Networks. *Proceedings of the 15th Conference of the European Chapter of the Association for Computational Linguistics: Volume 1, Long Papers* (pp. 1–10). Valencia: Association for Computational Linguistics.

Liu, Y., Han, T., Ma, S., Zhang, J., Yang, Y., Tian, J., … Ge, B. (2023). Summary of ChatGPT/GPT-4 Research and Perspective towards the Future of Large Language Models, Meta-Radiology, Vol. 1, Issue 2.

Liu, Y., Ott, M., Goyal, N., Du, J., Joshi, M., Chen, D., … Stoyanov, V. (2019). *RoBERTa: A Robustly Optimized BERT Pretraining Approach.* arXiv, abs/1907.11692.

Madaan, A., Tandon, N., Gupta, P., Hallinan, S., Gao, L., Wiegreffe, S., … Clark, P. (2023). *Self-Refine: Iterative Refinement with Self-Feedback.* Conference on Neural Information Processing Systems, New Orleans.

Marcus, G. (2020). *The Next Decade in AI: Four Steps towards Robust Artificial Intelligence.* arXiv:2002.06177.

Martin, J., & Jurafsky, D. (2014). *Speech and Language Processing An Introduction to Natural Language Processing, Computational Linguistics, and Speech Recognition.* Pearson.

Mialon, G., Dessì, R., Lomeli, M., Nalmpantis, C., & et al, R. P. (2023). *Augmented Language Models: a Survey.* arXiv:2302.07842.

Mohri, M., Rostamizadeh, A., & Talwalkar, A. (2018). *Foundations of Machine Learning.* The MIT Press.

Noori, B. (2021). Classification of Customer Reviews Using Machine Learning Algorithms. *Applied Artificial Intelligence, 35*, 567–588.

Nori, H., King, N., McKinney, S. M., Carignan, D., & Horvitz, E. (2023). *Capabilities of GPT-4 on Medical Challenge Problems.* arXiv:2303.13375.

OpenAI. (2023). GPT-4 Technical Report.

Park, J. S., O'Brien, J. C., Cai, C. J., Morris, M. R., Liang, P., & Bernstein, M. S. (2023). *Generative Agents: Interactive Simulacra of Human Behavior.* arXiv:2304.03442.

Peng, B., Li, C., He, P., Galley, M., & Gao, J. (2023). *Instruction Tuning with GPT-4.* arXiv:2304.03277.

Peters, M. E., Neumann, M., Iyyer, M., Gardner, M., Clark, C., Lee, K., & Zettlemoyer, L. (2018). *Deep Contextualized Word Representations.*arXiv, abs/1802.05365.

Phuong, M., & Hutter, M. (2022). *Formal Algorithms for Transformers.* DeepMind.

Prystawski, B., & Goodman, N. D. (2023). *Why think step-by-step? Reasoning emerges from the locality of experience.* arXiv:2304.03843.

Puchert, P., Poonam, P., van Onzenoodt, C., & Ropinski, T. (2023). *LLMMaps – A Visual Metaphor for Stratified Evaluation of Large Language Models.* arXiv:2304.00457.

Radford, A., Narasimhan, K., Salimans, T., & Sutskever, I. (2018). *Improving language understanding by generative pre-training.* arXiv:2012.11747, OpenAI.

Raffel, C., Shazeer, N., Roberts, A., Lee, K., Narang, S., Matena, M., ... Liu, P. J. (2019). *Exploring the Limits of Transfer Learning with a Unified Text-to-Text Transformer.* arXiv, abs/1910.10683.

Rohan A., Dai, A. M., Fira, O., M. J., Lepikhin, D., Passos, A., ... Wu, Y. (2023). *PaLM 2 Technical Report.* arxiv:2305.10403.

Rothman, D. (2022). *Transformers for Natural Language Processing: Build, Train, and Fine-Tune Deep Neural Network Architectures for NLP with Python, PyTorch, TensorFlow, BERT, and GPT-3.* Packt Publishing.

Russell, S. (2020). *Artificial Intelligence : A Modern Approach.* Pearson.

Sanderson, K. (2023). GPT-4 is here: what scientists think. *Nature, 773,* 615.

Sankar, C., Subramanian, S., Pal, C., Chandar, S., & Bengio, Y. (2019). Do Neural Dialog Systems Use the Conversation History Effectively? An Empirical Study. *Proceedings of the 57th Annual Meeting of the Association for Computational Linguistics,* ACL.

Shen, Y., Song, K., Tan, X., Li, D., Lu, W., & Zhuang, Y. (2023). *HuggingGPT: Solving AI Tasks with ChatGPT and its Friends in HuggingFace.* arXiv:2303.17580v1.

Sherstinsky, A. (2020). Fundamentals of Recurrent Neural Network (RNN) and Long Short-Term Memory (LSTM) network. *Physica D: Nonlinear Phenomena,* Vol. 404, 45, 67.

Stephanie Lin, J. H. (2022). *TruthfulQA: Measuring How Models Mimic Human Falsehoods.* arXiv:2109.07958.

Tafjord, P. C. (2018). *Think you have Solved Question Answering? Try ARC, the AI2 Reasoning Challenge.* arXiv:2109.07958.

Thoppilan, R., Freitas, D. D., Hall, J., Shazeer, N., Kulshreshtha, A., Cheng, H. T., ... Li, Y. (2022). *LaMDA: Language Models for Dialog Applications.* arXiv:2201.08239.

Touvron, H., Lavril, T., Izacard, G., Martinet, X., Lachaux, M.-A., Lacroix, T., ... Lample, G. (2023). *LLaMA: Open and Efficient Foundation Language Models.* arXiv:2302.13971.

Tunstall, L., von Werra, L., & Wolf, T. (2022). *Natural Language Processing XE "Natural Language Processing" with Transformers: Building Language Applications with Hugging Face.* O'Reilly Media.

Vasiliev, Y. (2020). *Natural Language Processing with Python and SpaCy: A Practical Introduction.* No Starch Press.

Vaswani, A., Shazeer, N., Parmar, N., Uszkoreit, J., Jones, L., Gomez, A. N., ... Polosukhin, I. (2017). *Attention Is All You Need.* arXiv, abs/1706.03762.

Wake, N., Kanehira, A., Sasabuchi, K., Takamatsu, J., & Ikeuchi, K. (2023). *ChatGPT Empowered Long-Step Robot Control in Various Environments: A Case Application.* arXiv:2304.03893.

Wang, Y., Kordi, Y., Mishra, S., Liu, A., Smith, N. A., Khashabi, D., & Hajishirzi, H. (2022). *Self-Instruct: Aligning Language Model with Self Generated Instructions.* arXiv:2304.03893.

Wang, F. Y. (2023). *Natural Language Reasoning, A Survey.* ACM Computing Surveys.

Wei, J., Wang, X., Schuurmans, D., Bosma, M., Ichter, B., Xia, F., ... Zhou, D. (2023). *Chain-of-Thought Prompting Elicits Reasoning in Large Language Models. NIPS '22: Proceedings of the 36th International Conference on Neural Information Processing Systems,* November 2022, 24824–24837.

Wies, N., Levine, Y., & Shashua, A. (2023). *The Learnability of In-Context Learning.* arXiv:2303.07895.

Wilmott, P. (2020). *Machine Learning: An Applied Mathematics Introduction*. Panda Ohana Publishing.

Wu, C.-S., Madotto, A., Hosseini-Asl, E., Xiong, C., Socher, R., & Fung, P. (2019). Transferable Multi-Domain State Generator for Task-Oriented Dialogue Systems. *Proceedings of the 57th Annual Meeting of the Association for Computational Linguistics*. ACL, Florence, Italy, 808–819.

Wu, S., Irsoy, O., Lu, S., Dabravolski, V., Dredze, M., Gehrmann, S., ... Mann, G. (2023). *BloombergGPT: A Large Language Model for Finance*. arXiv:2303.17564.

Xu, Y., Lee, H., Chen, D., Hechtman, B., Huang, Y., Joshi, R., ... Chen, Z. (2021). GSPMD: General and Scalable Parallelization for ML Computation Graphs. arXiv:2105.04663.

Yang, Q., Yu, Z., Dai, W., & Pan, S. (2020). *Transfer Learning*. Cambridge University Press.

Yang, Z., Dai, Z., Yang, Y., Carbonell, J. G., Salakhutdinov, R., & Le, Q. V. (2019). *XLNet: Generalized Autoregressive Pretraining for Language Understanding*. arXiv, abs/1906.08237.

Yu, F., Zhang, H., & Wang, B. (2023). *Natural Language Reasoning, A Survey*. arXiv:2303.14725.

Yue, Y. Z. (2023). *Meta-Transformer: A Unified Framework for Multimodal Learning*. arXiv:2307.10802.

Zhang, R., Han, J., Zhou, A., Hu, X., Yan, S., Lu, P., ... Qiao, Y. (2023). *LLaMA-Adapter: Efficient Fine-tuning of Language Models with Zero-init Attention*. arXiv:2303.16199.

Zhao, W. X., Zhou, K., Li, J., Tang, T., Wang, X., Hou, Y., ... Wen, J. R. (2023). *A Survey of Large Language Models*. arXiv:2303.18223.

Index

Printed in the United States
by Baker & Taylor Publisher Services

Printed in the United States
by Baker & Taylor Publisher Services